NOT SO DIFFERENT

NATHAN H. LENTS

NOT SO DIFFERENT

Finding Human Nature in Animals

Columbia University Press / New York

Columbia University Press
Publishers Since 1893
New York   Chichester, West Sussex
cup.columbia.edu

Library of Congress Cataloging-in-Publication Data

Names: Lents, Nathan H., author.
Title: Not so different : finding human nature in animals / Nathan H. Lents.
Description: New York : Columbia University Press, 2016. | Includes
    bibliographical references and index.
Identifiers: LCCN 2015039849| ISBN 9780231178327 (cloth : alk. paper) |
    ISBN 9780231541756 (e-book)
Subjects: LCSH: Animal behavior. | Psychology, Comparative.
Classification: LCC QL751 .L42 2016 | DDC 591.5—dc23
LC record available at http://lccn.loc.gov/2015039849

Columbia University Press books are printed on permanent
and durable acid-free paper.
Printed in the United States of America

COVER DESIGN: Milenda Nan Ok Lee

Para mi familia. Nora, "You're the BEST!"

CONTENTS

CONTENTS

ACKNOWLEDGMENTS

I HAVE TO BEGIN by acknowledging my mother, Judi, who has been telling me for almost two decades to write a book. I finally did and I hope you like it. As for my dad, Mike, you were the unwitting inspiration for this book. I'll tell you the story sometime. I also have to thank my always-encouraging husband Oscar, who read every word of this and endured me droning on about each topic endlessly, all while remaining supportive.

I would also like to thank the professional colleagues, friends actually, who read early samples and gave me crucial advice that shaped the course of the project in important ways. Evan Mandery and Richard Haw, you helped much more than you realize. Marc Bekoff, Joan Roughgarden, and Micchio Kaku, you were generous enough to consider my ideas and give advice and encouragement when I needed it most. I also want to thank my friends who read this book as examples of my target audience and gave me very helpful feedback regarding the tone, coverage, and detail. Lisa, Angela, and Oscar, I highly appreciate the effort you took to read the early drafts of this and the feedback you gave. You substantially helped me shape this book. I also want to acknowledge the support of the John Jay College Office for the Advancement of Research.

I also had two outstanding editors who gave this book the critical eye it needed and substantially improved my writing in the process. Heather

Falconer, your input early on helped me understand what it really means to write a book. The project would not have gotten out of the gate without your help. Tara VanTimmeren, this book would be a poorly organized, sloppily written, second-rate manuscript of interest to no one without your skilled work on it. (That last sentence is an example of how crappy my sentences are without your help.) Wendy Lochner from Columbia University Press has also been a fantastic acquisitions editor. Always patient and candid in the face of my relentless nagging and naïveté, she held my hand through this harrowing gauntlet of academic publishing.

And the biggest thanks goes to the students I've had the pleasure to learn with at John Jay College. I'm always embarrassed to be called your professor because I learn far more from you than you do from me. I have too many favorites to mention, but those that have "pushed back the frontiers of science" in my research lab deserve special mention because of all they've had to put up with from me. Kate, Zuley, Richard, Michael L., Abhi, Szilvia, Tamykah, Tetyana, Andrea, Rob, Andre, Derek, Stephania, Ana, James, Bridgit, Michael W., and Donovan: you inspire me daily. Now get out there and make me famous.

NOT SO DIFFERENT

INTRODUCTION

Emotions, Drives, and the Brain

WHEN WE DISCUSS the emotions, behavioral drives, and even thoughts of humans and other animals, we are inevitably caught in the complex web of the relationship between the brain and the body. Our outward behavior is the result of how our emotions, drives, and reasoning move us to act. Although all animals are born with genetically directed tendencies toward certain urges, desires, and emotional reactions, those drives and emotions gradually take shape as our brains and bodies interact with our environment, first in utero, then in childhood, and up through adulthood.[1] Particularly with humans, we also develop the ability to reason and this, too, affects our decision making and ultimately our behavior—the way that our brain moves our body to act.

For both animals and humans, the brain is at the root of behavior, integrating all the various inputs and selecting and organizing all the outputs. It is not, however, a simple input-output system. It operates within the parameters of its capability, of course, and does so according to certain genetic programming, but it is also guided by the imprints of past experience. Day after day, our behavior is the result of complicated interactions of a staggering array of factors, including personality, past experience, and genetics.

The thesis of this book is that underneath even our most complex behaviors are some rather simple, genetically encoded predispositions that we share with many other animals. This is not to say that human behavior can be reduced to simple urges, but rather that much of it is situated atop some basic behavioral scaffolding that is shared with most other social animals. This behavioral scaffolding is tweaked in unique ways across species and even among individuals within a species, but some general trends reveal some common tendencies.

In this book, I discuss experiments that have revealed features of animal behavior that are strikingly similar to human behavior. These similarities, as far as I can tell, can only be explained by acknowledging that human and animal brains run some of the same behavioral programs. Some see all of this complex animal behavior and conclude that the inner experience of animals must be much more involved than we tend to appreciate. An equally important conclusion is that many seemingly bizarre and counterintuitive human behaviors can be understood by analyzing the underlying pursuits and desires, which may actually be rather simple. The complexity of the human brain may be a smoke screen, clouding the simplicity underneath. Either way, the point here is that the gulf between humans and other animals is not as wide as common wisdom has traditionally held.

Even if we view animals as mindless automatons, like robot bodies controlled by computer brains, still, a computer must run programs. The thesis of this book is that the suite of programs that underlies animal behavior is remarkably similar to that which underlies human behavior. It could be that human behavior seems more complex because of one key difference: our abilities in advanced reasoning, which we all agree exceed those of other animals. This advanced reasoning adds another layer of input into our behavior as we contemplate decisions and their consequences. Nevertheless, it is my contention, and that of many other scientists who study human and animal behavior, that the underlying behavioral programs of humans and animals are quite similar.[2]

This brings me to the reason I wrote this book. It is my belief that learning about animal behavior can teach us about human behavior. We can run experiments and make observations with animals that would be impossible to do with humans, and what we learn from animals can inform how we understand our fellow humans. When we make progress in deciphering the root causes of our desires, instincts, urges, and behaviors, we can better respond to them.

Understanding where sibling rivalry comes from can help us disarm it and thus get along with our siblings better. Understanding the biological basis of grief can help us recover from our own grief as well as help others to do so. Understanding that humans have a moral foundation built into us through our history as social mammals can help us discover ways to build a more moral society, regardless of religious, national, and ethnic differences. In short, there is a lot to gain by understanding the origin of our behaviors.

BEHAVIORS ARE BORNE OF GENETICS

The way that we often study animal behavioral patterns is to place animals in specific circumstances and watch. Underneath this approach is the assumption that the behavior we observe is at least partially programmed by genetics, and these inborn, preprogrammed aspects of behavior are the focus of this book. Of course, animals have some learned behaviors also, both on their own and resulting from instruction from others, but we usually know when that is the case, and we can control for that in our experiments. For example, we can raise laboratory animals in a way that prevents their learning of certain behaviors. This is not as easy with animals in the wild, but again, we can usually find out if behaviors are learned or genetic if we watch closely enough.

Why does this matter? When it comes to understanding how our genes work, most people have no trouble accepting that genes encode things like eye color, height, skin tone, and even susceptibility to certain diseases. However, it is not so obvious how a complex emotional state—jealousy, for example—could result from genes inherited from the sperm and egg of our parents. It seems unfathomable that mere DNA and protein could control emotion and behavior, but what else could control them? Anyone who has ever seen a brood of ducklings following their mother has borne witness to the power of genetically programmed instinct in directing behavior. Anyone who has seen a duckling that hatched in the presence of a dog, instead of its mother, has seen it even more clearly.

When I was growing up, my family owned an Old English Sheepdog named Daisy. I am quite certain that Daisy had never seen a farm animal in her life. She was completely oblivious to the selection that had taken place in her ancestors, which led to the incorporation of herding instincts into her genetic complement. I will never forget the first time Daisy saw a group of toddlers playing in our living room. After briefly assessing the situation, she began to circle them slowly and regularly, switching directions and pacing. If one of the toddlers started to drift out of the imaginary circle she had drawn for them, she gently nudged him or her back inside it. She instinctively herded these toddlers, much to their annoyance, and did so with efficiency and focus. Her ancestors would be proud! She just knew what to do and how to do it. If this behavior were not coded in her genes, where else did it come from?

Behaviors can be inherited. Indeed, many of them are. Of course, humans have a more intricate psychology than other animals, so the interaction between our genetic programming and our environment makes things very complicated. I do not think Freud and Jung had too much to say about the psyche of cows. However, underneath the complexity are some basic genetic programs, and, to be fair, environment affects the development of animal behaviors from their genetic programming as well.

Human behaviors, complex though they are, emerge from much simpler behavioral programming that we inherited from our ancestors and that we share with other animals. This programming continues to define major aspects of who we are and how we act. If you do not believe me, keep reading. In the pages ahead, you will not only read about things we have learned from animal studies but things we have learned from studies on humans as well. We are not so different from our animal cousins.

NATURAL SELECTION IS THE ENGINE
OF EVOLUTION

This book hinges on the scientific truth that all life on Earth is related through common ancestry—all life, humans included. We are descended from ancestors that we share with other species. Our closest relatives are the chimpanzees and bonobos, often called pygmy chimpanzees. Those two species are closest to us because we have a more recent common ancestor with them than we do with any other species. The common ancestors of humans and the two species of chimps lived just six million years ago and did not perfectly resemble either chimps or humans. That species is now extinct, but some members gave rise to our lineage, and others gave rise to the two chimp species.

Evolution refers to the gradual change of lineages of organisms over time. These changes occur through a variety of means, and natural selection is just one of them. It is the most famous force, of course, and the one first championed by Charles Darwin and Alfred Wallace in the 1860s. However, the concept that lineages of species evolve over time was already widely accepted among biologists in the time of Darwin.

Natural selection, despite being the unifying theme of biology, is still commonly misunderstood. It is often summed up as "survival of the fittest." This phrase only captures natural selection if you are using "fit" in a very precise way—the biological way, in which it means "leaves lots of offspring that are fertile and prolific." In common language, "fit" means to be in good shape, to be healthy, to be successful, and so forth. In biology, it refers only to reproductive success. Part of being reproductively successful is living long enough to reproduce, so of course one has to be healthy to accomplish this. Being faster, smarter, stronger, and hardier are all very good ways to survive, attract a mate, and produce lots of offspring, so, in general, we have seen that those traits have been selected positively in many species. However, those traits are not enough. No matter how healthy, strong, or smart you are, if all of your children are killed by a sneaky rival, you will not contribute to the next generation.

In order to fully appreciate evolution, several common misconceptions must be avoided. First, individuals do not—indeed, cannot—adapt, biologically speaking. In common usage, the word "adapt" means to change or adjust to new circumstances. In biology, adaptation refers to a specific kind of change: genetic change. You cannot alter your genetics. You are stuck with what you are born with, and any changes that you make to your body cannot be passed on. Although individuals cannot adapt, populations can, over time, through the differential survival of some members over others.

If a sudden drought occurs that kills 95 percent of the individuals in a certain flock of birds, it is possible that the ones that survive have some sort of natural variation that helped them. It may be a natural thirst resistance or the strength to fight off other birds for precious water. Or perhaps it is fatty deposits that insulate internal water from evaporation. The point is that the difference will be abundant or even fixed in future generations. That is adaptation.

The second-most common misconception about natural selection is that the selection itself somehow creates the new adaptations that emerge. For example, many people might think that leaves on tall trees caused giraffes to grow long necks. Proto-giraffes had small differences in neck length, like all animals do, and those with longer necks survived and reproduced a little bit better than those with short ones. Each generation brought more variation, and the selection continued, but tall trees did not somehow cause the growth in the giraffes' necks. The variation was already there, and the tall trees just helped to select winners and losers from the pool of hungry giraffes.

Similarly, antibiotics do not create antibiotic resistance; they just select for it. The natural variation when it comes to resistance already exists. Some bacteria are naturally a little more resistant than others. With the widespread irresponsible use of antibiotics, we are selecting, again and again, those with more resistance. Over time, with more spontaneous variation emerging in the bacteria and more selection being applied by antibiotics, the population of bacteria becomes more and more resistant. However, no individual bacterium adapts or evolves, and the antibiotics themselves do not create the spontaneous resistance. The antibiotics just ensure that those bacteria with the resistance will outcompete those without it, thus rendering all descendent bacteria resistant.

Living things have a tendency to create variation. Whether because of errors in DNA replication or the shuffling of genes during sexual reproduction, organisms are diversity-generating machines. Nature then acts on that variation. The features that happen to best suit the environment at that particular time and place will naturally persist because of the success that they confer upon their host. However, the creation of more variation will continue. The neck of the giraffe evolved to become very long over millions of years of incremental change and persistent directional selection.

Another misconception is that living things are perfectly adapted for the environments they live in. Of course, all living things are well adapted, or else they would quickly become extinct. However, nothing is ever perfectly suited to its surroundings—at least, not for very long. Habitats are continuously changing in a variety of ways. Besides, were a species to become perfectly adapted, it would overpopulate rapidly, eradicate its food source, and boom: not so well adapted anymore.

The harsh reality is that, for pretty much all species at all times in history, the struggle for survival has been so brutal that the vast majority of individuals have fallen victim to predators, competitors, injuries, or infectious diseases long before they could leave any offspring. In fact, this was one of Darwin's key observations that changed his thinking about life. Every single species has a tendency to produce far more offspring than can possibly be supported by the environment. Darwin watched pairs of finches create nests and nurture four or five fledglings every single year for their entire lifespan of ten to fifteen years. And yet, the population of the finches remained stable. This meant that most baby finches did not survive to become parents themselves.

Oak trees drop thousands of acorns every year. Dandelions spread thousands of seedpods. Fish lay and fertilize hundreds of thousands of eggs. Wolves have litters with dozens of pups. If every acorn grew into an oak, within a few hundred years, the Earth would be nothing but oak trees and acorns. If every baby male gorilla grew up to establish his own harem, there would be room for little else in this world but gorillas. No matter how romantically we look upon the natural world around us, it is actually an exceedingly cruel place to live. All species overproduce, and most individuals die in the resulting struggle. This is the survival of the fittest of which Darwin spoke.

So intense is the struggle for survival that species must constantly evolve and adapt, not so they can thrive but merely so they can avoid extinction. For any given species, its predators (or herbivores, in regard to plants) are also evolving, as are its competitors and parasites. Its food source, whether plant or animal, is likely evolving to avoid being eaten. Even pathogens evolve ways to defeat immune systems, and this may be the most intense threat of all. To survive is to change, and there is no rest for the weary. Biologists call this the Red Queen phenomenon after the Alice in Wonderland story in which the Red Queen must run faster and faster in order to stay in the same place.[3] Survival on planet Earth is rather like a treadmill suspended over a snake pit. No matter how fast you run, you never get further from the snakes—but if you do not run at all, you will die.

This leads to yet another common misconception about natural selection. Beginning with Darwin and continuing today, most of the public and many scientists themselves have concentrated only on the competitive side of nature. In this view, any two members of the same species have no relationship with each other except as competitors. While this is true in many species, another evolutionary strategy has also emerged, especially in our own lineage: social cooperation. Communal living has allowed for the emergence of cooperative resource acquisition and sharing, common defense, division of labor, and other survival strategies that promote trust and cohesion among members of a group.[4] This does not completely eliminate competition among members, but it does mean that they can work together for mutual success. Much of this book is devoted to human phenomena that have evolved from the same cooperative features found in other social animals.

The evolution of cooperation is still an actively debated concept in the world of evolutionary biology, in part because the savage conditions in which most species exist seem to favor selfishness over altruism. It is very difficult to see a way that group cooperation could emerge and dominate before selfishness beats it back. This is why most biologists have withheld their support for so-called group selection, which is the idea that a trait could be selected by nature through the success it provides to the group rather than to individuals. However, more recently, a combination of kin selection (helping one's close relatives and thus one's own gene pool) and reciprocal altruism (returning favors) has proven to be a successful strategy, even better than selfishness, in simulations and laboratory experiments.[5]

What this means is that cooperative behaviors, in certain instances, can actually help an individual succeed, provided that pack sizes are small and the animals have good enough memories to recall who is naughty and who is nice. Small group size works to foster cooperation because, in that scenario, just about everyone who comes in contact will be either close relatives or acquaintances who will likely come in contact again.

The fossil and archaeological evidence overwhelmingly reveals that, up until around eleven thousand years ago, humans lived in very small communities.[6] Only the invention of agriculture and livestock allowed bands and tribes to establish permanent settlements that then grew to form small villages and hamlets. In other words, for nearly all of the two-hundred-thousand-year history of Homo sapiens, living conditions were small, closely knit communities in which most people were related and everyone knew everyone else. In other words, it was ideal for natural selection to nurture the development of cooperation.

Truth be told, the trend goes back much further than that. Just about all known primate species live in small bands. These bands are highly social and, in most cases, stratified into dominance hierarchies. Many primates build social alliances, divide labor, share resources, and raise their young communally. In other words, they cooperate. Primates are not alone in this, either. There is abundant evidence that many birds and mammals exhibit cooperative behaviors.

The brilliant biologist and thinker Lynn Margulis put it this way: "Life did not take over the globe by combat, but by networking."[7] Although she was speaking to how eukaryotes (plants, animals, fungi, and protists) came to dominate over prokaryotes (bacteria and Archaea), the same truth applies to humans. It is difficult to imagine that humans would have come to dominate the Earth without our tendency to work well together.

EXTREME SELECTION LEADS TO RAPID CHANGE

Despite the overwhelming evidence that natural selection is the primary vehicle for adaptive evolutionary change, doubt persists among the general population regarding the role of natural selection in the development of the modern human experience. This doubt is understandable because there is so much about our psychological and sociological features that are just phenomenally different from anything seen in animals. How could we have become so different so quickly?

There are several answers to this. First, so-called human nature is not so different from behavioral features found in other animals. That is what this book is all about. Second, much of what we think of as uniquely human comes from constructed culture. The ability to construct culture probably came from a tiny biological advance in the ability of the human brain. This had profound impacts because it allowed the development of language. Language, it is argued, then allowed us to become capable of complex thought, and everything else flowed from that.[8] This could have happened overnight, almost literally speaking, like a switch that turned the cognitive light on in our ancestors, opening up the rich introspective inner experience that we know as human consciousness.

Third, it is important to remember that for about six million years, our lineage has been evolving separately from any other species alive today. That is a long time—several hundred thousand generations. During that time, our family tree has grown many branches. We do not know how cognitively advanced the lost branches really were. It is difficult to tell from skeletons and possible artifacts. Who is to say that if we resurrected a Homo erectus and brought her up in our modern world, she would not behave a lot like we do? This view also discounts how rapidly a strong selective pressure can induce evolutionary change.

This last point brings us to yet another misconception about the pace of evolutionary change. Everyone knows that evolutionary change, on the whole, is very slow. However, it is almost never the case that evolutionary change occurs at a steady, incremental pace. From the fossil record, we know that species often go long periods of time with little or no observable change. Then, quite suddenly, they will undergo an adaptive change. One of the most well-established explanations for these alternating cycles of stasis and change is called punctuated equilibrium.[9] When we ask how humans could have possibly accumulated so many big differences from other animals in such a short amount of time, we must keep in mind that punctuated equilibrium predicts that species-level evolution can actually occur quite rapidly in certain circumstances of intense selection. For example, a 2014 study reported that lizards called anoles evolved measurable anatomical adaptations in just fifteen years.[10]

"But wait," you might ask. "You just said that survival is always a struggle! How are the moments of so-called intense selection any different from the rest of the time if the struggle to survive and reproduce is always very intense?" Aha! Therein is the key to understanding rapid evolutionary change. It is true that surviving is always a struggle and that most individuals will perish before reproducing. What is different about periods of rapid change is that the selection is persistent and directional when it comes to a certain trait or group of related traits.

Imagine life for a European hare. These animals reproduce like, well, rabbits. They have fast metabolisms and are constantly eating and moving. They grow and develop rapidly and rarely rest. They can complete a full reproductive cycle in a couple of months, and they have large litter sizes. A single pair of hares can give rise to hundreds more in less than a year. And yet, in Europe, their numbers have been stable for hundreds of years and may actually be declining today. In other words, very few hares that are born will leave any children of their own, let alone grandchildren.

Does the European hare's intense struggle to survive represent intense selection? Almost certainly not. Why not? Because there does not appear to be any net directional selection of any specific traits. The hares are perishing for a million different reasons. Some cannot find shelter in winter and then freeze. Some fail to hear, see, or outrun a predator. Some eat a poisonous vegetable, while others chew an electrical wire. Some hares cannot attract a mate and reach the end of their short lifespans as frustrated virgins, while some have low sperm counts and lose out to more virile males. We could go on for hours and still not cover all the ways that hares meet their doom.

There is danger on all sides for these little guys, which is probably why they evolved their prolific reproduction in the first place. That may be the only way they can keep their numbers high enough to avoid extinction. The point is that this kind of selective pressure is too scattered to lead to any specific adaptations. There is no net selective force—except maybe just for general health, survival, and reproductive prowess—and so the species will just stay as it is. The constant general selection will keep the species on its toes, for sure, but this is not a recipe for developing new adaptations.

Now compare this with what might happen to European hares when a new predator is introduced to their habitat. Suddenly, there is a new sheriff in town, and there will be a whole lot of a particular manner of death. Under these circumstances, the hares may experience selection that does have a specific direction. Assuming they are not hunted to extinction first, the hares will develop some adaptive defense against the predator. It could be speed, camouflage, better hearing or eyesight, the ability to climb bushes or burrow underground, or maybe even an unpalatable taste in their flesh. (There are species of tree frogs that evolved poison in their skin to punish their predators and train them to avoid eating them.) The point is that adaptations tend to appear suddenly in response to a very specific and overwhelming threat. This is the kind of intense selection we are talking about. Remember, though, that this selection does not create the adaptation itself. The hares have only their natural variation to work with.

The selective pressure applied by a new predator is a form of negative selection. There is positive selection as well. This could occur, for example, when our hares expand into a new territory with an abundant new food source. Perhaps their teeth are not very well suited for the new food. This would represent intense positive selection because the few members of the population that are naturally gifted with slightly better-shaped teeth for the new food source would have an enormous advantage and would begin to flourish. They and their offspring would overwhelm the others, and over time the shape of the teeth in this population of hare would change. If this population was separated from the original parent population long enough, it could become a new species, and the two could drift farther and farther apart as their environments changed.

THE RAPID RISE OF H. SAPIENS SAPIENS

What does all of this have to do with human and animal behavior? Over the last six million years and particularly over the last one million, the various species of protohumans experienced intense positive selective pressure for cognitive abilities and intelligence. The stage was set. Apes and hominoids had been evolving more intricate and tight-knit social communities. They had also been adapting to become increasingly intelligent, likely because of that sociality. The habitat of the African savannah, with its lush plant and animal biodiversity interrupted by frequent and unforgiving changes in climate, was an environment that particularly favored cleverness. These were ideal conditions for rapid evolutionary change. That the human lineage developed unprecedented mental abilities is altogether unsurprising.

In fact, the evolution of Homo sapiens from earlier species of Homo and, even before that, from species of apes, is a story of how our evolution focused almost solely on increasing intelligence at the expense of just about everything else. Compared with other great apes, we are incredibly weak and slow. Our eyesight diminishes very commonly and very early. Our hearing is not that great, and is certainly worse than that of other apes. Our reproduction is horribly inefficient and often deadly. We have lost most of the hair that provided several benefits in addition to the obvious insulation from cold air. Speaking only of physical ability and physiological functions, there is nothing that we do better than other apes, and there are many things that we do worse. It is true that we walk upright while other apes do not, but we are still much slower on our two limbs than chimps and gorillas are on four. We sacrificed all kinds of physical abilities in our relentless pursuit of mental ones. That was the trade-off.

Because humans are now the most abundant mammal on the planet, it is somewhat hard to imagine us ever going extinct. However, that is exactly what almost happened—many times, in fact. From the fossil record and from DNA analysis, we can tell that our ancestors have teetered on the verge of extinction, dwindling to very small numbers countless times. In addition, there are many lineages of hominids that did go extinct. Since the split between our ancestors and those of the chimps, our lineage has not been a single line of gradual change. Evolution never works that way. Instead, many branches broke off from each other and developed branches of their own. There were at least three or four different species of hominids living simultaneously for most of the past five million years. Of all these branches, only one survived until today: ours.

As recently as forty-two thousand years ago, two other branches of Homo persisted as well—the Neanderthals and the Denisovans—and Homo erectus persisted in Java until no more than seventy thousand years ago, well after modern humans had entered the scene.[11] These other human species were almost—though not quite—like us, but they were not our ancestors. They were our cousins, and we coexisted with them for at least one hundred thousand years, although mostly in nonoverlapping territories. Although modern humans occasionally interbred with both Neanderthals and Denisovans, our species did not simply merge with theirs. Since we may not have found all the fossils (yet), who knows how many other hominid species came and went?

What can we make from all this extinction? Life was rough for these hominids. Just surviving the day was a brutal struggle. Suffering, starvation, disease, and death were all around. It was a bleak and nearly hopeless existence. Somehow, enough protohumans survived each catastrophe, while continuing to evolve, that eventually our ancestors made the final great leap forward into what we call behavioral modernity. This was sparked by the emergence of language. Since then, we have flourished and spread to all corners of the globe, coming to dominate and transform every environment we invade.

Although we seem so drastically different from any other life form, we really are not. If you consider our anatomy, physiology, and biochemistry, there is little to distinguish us from other mammals. Our cells work exactly the same way; our tissues cooperate the same way; our blood circulates the same way. The only real differences are in our brains, and even those are not that stark. If you compare the dissected brain of a human with that of a chimpanzee or even a dog, cow, or sheep, you will see that they are not that different. Human brains are bigger, pound for pound, but most of that additional mass is not due to having more neurons but to having more connections between neurons. We do not have that much more gray matter than other animals; we have a lot more white matter.

The thesis of this book is that all of the impressive human cognitive abilities evolved in ancestor species that already had an extensive palette of emotional states. In order to understand how our ancestors made the jump from animal behavior to human psychology, we must first recognize that the distance of that evolutionary jump is not as great as it seems—and six million years is plenty of time to have made it. I argue here, as others have, that the gulf that differentiates the human emotional experience from the animal one is due solely to advances in the ability of the human brain. In this way, I do not think we invented any new emotional experiences at all. Who knows? We may have actually lost some emotional qualia. Rather, I think our impressive abilities in calculation, memory, and introspection have merely allowed us to contemplate our emotional states and attempt to understand them.

I think both humans and chimpanzees feel love; the only difference is that humans write sonnets about it. I think both humans and dolphins practice fair play, but only humans enact laws to govern it. I think both humans and elephants experience grief, but only humans seek professional counseling to cope with it.

If you do not believe me, keep reading.

FURTHER READING

Balcombe, Jonathan. *Second Nature: The Inner Lives of Animals*. New York: Macmillan, 2010.

Bekoff, Marc. *The Emotional Lives of Animals: A Leading Scientist Explores Animal Joy, Sorrow, and Empathy—and Why They Matter*. Novato, Calif.: New World Library, 2010.

Dawkins, Richard. *The Greatest Show on Earth: The Evidence for Evolution*. New York: Simon and Schuster, 2009.

Pinker, Steven. *How the Mind Works*. New York: Norton, 1997.

1

WHY DO WE PLAY?

WHAT COULD BE MORE HUMAN than recreation? A game of chess, a mystery novel, a video game, collecting stamps, playing tennis, a night of drunken debauchery. There are an infinite number of ways that humans have fun, and they all seem very human-specific. Who can imagine raccoons sitting around playing a game of Texas Hold'em? While dolphins are undoubtedly intelligent, I am pretty sure they would not play chess for kicks, even if we could teach them how. So, do animals have fun? Is there an equivalent of recreation in the animal world? Hold that thought for a second. Before we look at animals having fun, we will explore why humans have fun.

At first glance, there does not seem to be any evolutionary advantage to having fun. How could it have ever been advantageous for humans to be so preoccupied with games and play? In fact, during the long and dark pre-agricultural period in which the very existence of our species hung by a thread and every day was a struggle, would playing have been a detriment to survival? In humans, as well as other animals, time spent playing is time not spent looking for food. Play could be a distraction from looking out for predators. Play could cause needless injuries—even traumatic death. Viewed from the lens of species survival, the drive to play seems like it would be a huge disadvantage. But yet, all humans in all cultures like to play.

The universality of play indicates that it is an innate feature of humanity. It had to be present in our ancestors, even while they were struggling during those dark times. Furthermore, playing is not exactly a minor part of our lives. It is a big and important part of who we are. This behavior cannot simply be an evolutionary side effect or a genetic accident. We are not talking about a tiny appendix in our abdomen. Play is a huge part of the human experience, especially among the young, and thus there must be value in it.

Just like organs or tissues, behaviors will be carefully honed over time by natural selection. Behaviors that enhance survival and/or reproduction of individuals will develop through the generations and become part of the innate "nature" of the species. Such as the suckling of all infant mammals, many of our behavioral urges are driven purely by instinct. On the other hand, behaviors that detract from survival or reproduction will quickly be eliminated through the occasional death of those inclined to perform them, provided there is some genetic influence on the behavior. Further, the more time that an animal spends engaging in a behavior, the more certain we are that it plays an important role in survival or reproduction.

Similar to seemingly useless anatomy, one could reason that some occasional behaviors are mere flukes. Maybe they are carryovers from a previous environment in which such behavior, or some earlier form of it, was necessary. After all, natural selection does not work overnight. Just because a behavioral drive is no longer needed does not mean that it will disappear from a species instantly. It takes time and negative selection. Individuals have to die and take these no-longer-helpful behaviors with them in order for the species to evolve away from them. This cannot be the case with play, even in humans. All humans in all cultures have a drive to play, and we spend a lot of time doing it. With this in mind, it is simply not possible that recreation is not somehow important for human health, survival, or reproduction. But how?

Now that we have properly framed the question, we can go back to animals as a starting point. Do animals play? Right off the bat, any dog owner will tell us that of course they do. They play fetch, they do tricks, they play tug-of-war. Dogs chase each other, they wrestle, they roll on the ground together. Some dogs like to swim; some enjoy jumping into the water from a boat or a dock. So the answer is yes, dogs are playful animals.

Actually, we cannot let ourselves off that easy. Dogs may be a special case because, through their unique breeding, they have a strong, innate desire to work and please their human companions.[1] Stronger in some breeds than others, this is a well-documented feature of dogs that is unique to them. It is a result of thousands of years of their evolving alongside humans, as well as selective breeding by those same humans. Dogs really are programmed to "enjoy" working with and for humans, and the only required reward is praise and acceptance. Yes, dogs will also do tricks and perform tasks for food or privileges, but with the proper training, the natural desire for dogs to be accepted and praised by their masters is enough to entice many to perform virtually any task or behavior. In fact, one study has shown that dogs are physiologically "aroused" when they perform trained tasks.[2] The point here is that we cannot take the easy road and say that animals know what "fun" is just because dogs seem to play with their human companions. We have to dig a little deeper.

What about when dogs play with each other? Undoubtedly, dogs are naturally given to wrestling with and chasing each other. However, rather than purely for fun, these behaviors could be attempts to establish dominance within a social context and a means to establish trust and familiarity. This does not necessarily mean that these behaviors are not also fun for the dogs, but it does mean that we cannot claim that they are, per se, evidence that animals play just for the fun of it, like humans do. That is the kind of play that we are talking about—just plain old having fun. That is what we humans do. We just have fun. For humans, play is for play's sake.

Or is it? Maybe our definition of play is the sticking point. If we think of play as doing something enjoyable that has no other purpose, then of course we will not find another purpose—we have made *not* having a purpose part of the definition. Instead, we need to keep an open mind about hidden purposes of play. We play because it is fun, but it may also be serving other purposes. After all, if we discover biological benefits to playing, then it really is not "just for the fun of it." It only seems that way. Perhaps the example of dogs playing together really is a clue to the function of play for other species, including us. Perhaps the secret to understanding the function of play is the realization that playful acts can be fun and serve some other purpose for the species.

DEFINING PLAY

Before going further, we must supply a definition of play, which is more difficult than one might think. Any attempt at a simple definition fails rather quickly. One problem is that any subjective features are not helpful. Words like "fun" and "enjoy" are hard to define in other species and could also be applied to things that are clearly not play. For example, eating, sex, and even scratching an itch are things that we enjoy and often have fun while doing, but they are very different behavioral phenomena than things like low-stakes games, sports, make-believe, and so on. We need a definition that is more restrictive than simply things we enjoy or things we do that are fun. Another problem is that a definition along the lines of "serving no purpose" dooms our whole discussion because we are looking for what the benefits of play might be.

In the comprehensive and erudite book *The Genesis of Animal Play*, reptile behaviorist Gordon Burghardt spends well over sixty pages discussing the problems of defining what play is.[3] He demolishes most prior attempts to define play by pointing out how the definitions are either overly broad or overly restrictive. Instead, he offers a framework of the five principle features of play and argues that all five standards must be met, at least minimally, in order to label a behavior as play. The five features are:

1. The performance of [play] behavior is not fully functional in the form or context in which it is expressed; that is, it includes elements, or is directed toward stimuli, that do not [directly] contribute to current survival.

2. The [play] behavior is spontaneous, voluntary, intentional, pleasurable, rewarding, reinforcing, or autotelic ("done for its own sake").

3. [Play] differs from the "serious" performance of ethotypic behavior structurally or temporally in at least one respect: it is incomplete (generally through inhibited or dropped final elements), exaggerated, awkward, or precocious; or it involves behavior patterns with modified form, sequencing, or targeting.

4. The behavior is performed repeatedly in a similar, but not rigidly stereotyped, form during at least a portion of the animal's ontogeny.

5. The behavior is initiated when an animal is adequately fed, healthy, and free from stress (e.g., predator threat, harsh microclimate, social instability) or intense competing systems (e.g., feeding, mating, predator avoidance). In other words, the animal is in a "relaxed field."[4]

As you can see, defining play is serious business for the scientists that study it. The complexity of this definition reflects the complexity of play forms found in humans and other animals. Notice how many times the word "or" appears. This is to ensure that our criteria for play capture the rich diversity of play found throughout the animal kingdom. However, the requirement that all five conditions are met, in at least one aspect, ensures that nonplay behaviors are not counted as play.

Burghardt's list is a good example of how scientists attempt to achieve objectivity while studying something that is inherently subjective. The study of animal behavior constantly struggles to develop and apply objective measures for their work, which tend to change over time (hopefully for the better), and the literature on animal play reveals constantly evolving standards and approaches.

Lucky for us, we can get away with a deeply subjective definition of play for the purpose of our discussion. "You know it when you see it" will work just fine. While this is hardly scientific, it is good enough for this chapter because I will only be discussing examples and modes of play that meet the listed criteria and are accepted as play by scientists. If you hunger for the more objective and rigorous analysis of animal play, I urge you to read

Professor Burghardt's book. It is a dense text written for specialists, but at seven-hundred-plus pages, it is the most historically and scientifically comprehensive discussion of animal play out there.

ENJOYMENT DRIVES PLAY

Now that we have covered the scientific definition of play, I will discuss play as we know it: that which is fun. To be a little careful, we can add, "that which is fun but not something else like eating or sex." The point is that the essential element is that we enjoy play, which is what makes it play. But "fun" and "play" are not synonyms. Play is the behavior itself. Fun is the feeling we get when we do it. In all her creativity, nature has joined the two in order to drive us to do the behavior, just like sex.

Nature's way of nudging animals to do certain things is to make them enjoy those things. Enjoyment is the feeling of pleasure in our brains caused by the release of neurotransmitters, which is triggered by satisfying the urge to do some behavior.[5] All animals have a drive to eat and to drink; it is called hunger and thirst. As we will discuss later, we also have a sex drive. Our brains drive us toward these behaviors through the feelings of pleasure that we feel when we do them. Pleasure is the reward for performing some behavior that is important for survival or reproduction. The fact that we enjoy playing tells us that we have been hardwired to do it.

What are pleasure, fun, and happiness? Again, this could take us down a difficult road, but it is sufficient for us to say that neuroscientists have known for some time that there are chemicals called neurotransmitters that are released in the brain when we experience the feelings of joy and happiness. Many hormones have been implicated in pleasure and the mood we know as happiness, including dopamine, serotonin, norepinephrine, prolactin, oxytocin, and endogenous opioids.[6] Different fun or enjoyable activities will cause the release of different combinations of these and other neurotransmitters, and it is a complicated mix.

Through the genius of evolution, these neurotransmitter-driven feelings have evolved as reward systems that underpin our natural drives, urges, and cravings. How exactly the "feeling" of pleasure really came about is something of a mystery, but one thing that is very clear is that the reward system is tremendously similar in humans and our animal relatives. We know this because of animal models of reward and addiction and also because of our ability to measure the release of these neurotransmitters in real time, both in humans and in animals. From this, we know that these pleasure-reward systems can powerfully affect us and drive us toward certain behaviors. We play because we enjoy the neurotransmitter-induced feeling that it gives us when we do. The more interesting question is, why have mammals evolved to enjoy these play behaviors in the first place? Why does nature drive us to play?

PLAY FOR ESTABLISHING SOCIAL RANKING

We mentioned playful dogs earlier, so I will start there. In the case of dogs, wrestling and chasing and rolling around on the ground is how they establish dominance within the social hierarchy in their packs.[7] Remember that dogs are descended from wolves and thus retain much of their pack-based social interactions. The survival advantage of this kind of play now becomes clear. Wrestling and chasing are much safer ways to establish dominance than outright fighting. If the dogs were to actually fight in order to establish dominance, the loser could end up dead, and even the winner would likely sustain some injuries. By playing instead, both dogs benefit.

Indeed, some mammal and bird species employ brutal fighting in order to establish dominance, and they suffer losses when they do. Worse, this has led to the bizarre and costly adaptation of antlers and horns in some species, used primarily for fighting among themselves.

How do dogs know when they are playing versus when they are really fighting? As much as it may look like fighting, dogs actually communicate very clearly to each other when they are playing.[8] Play begins with an invitation called a play bow. We have all seen a dog do this: they leave their hind legs straight, and with their backside high in the air, they lower their heads and forelegs. This is an invitation to play or, more forcefully, a warning that "play is about to commence; here I come!"

Along with the bow, the tail is held high in the air and wags back and forth before and during the wrestling. This is a way to signal nonthreatening social interaction. You can read a lot about a dog's emotional state from her tail. High and wagging is a sign of pleasure and play, while a tail held low implies a dog that is scared, threatened, or on edge. Because tails are how dogs and wolves communicate with each other, we can interpret the signs as well.[9]

Another way that you can tell the difference between playing and fighting is that the dogs do not typically bite very hard during play. Dogs have powerful jaws and can produce a deep puncture wound with just one quick bite. And yet, playing dogs rarely get hurt. This is where the communication and the rules come in.

With this in mind, it seems entirely plausible that, in dogs and wolves, play is a safe alternative to fighting for working out social aggression, and that is the purpose we are looking for! Dogs are programmed to enjoy wrestling with other dogs because that drive helps to safely achieve social harmony. Although humans (except adolescent boys) do not seem to establish social hierarchies through wrestling, this provides a nice model for how to discover why humans play. We need to examine how humans play and then ask what it might accomplish other than simply "having fun."

PLAY FOR LEARNING SOCIAL RULES

One hypothesis that has emerged is that play is the means by which many social animals learn various rules about which behaviors are acceptable and which are not. This has been proposed most confidently for wolves, which would likely extend to dogs as well.[10] The idea is that, through play, animals probe for boundaries. They act out in various ways and indulge their natural impulses, but as they are "corrected" by the social environment, they learn how to behave. This discussion actually delves into the issues of rules, punishment, fairness, and justice. (The notion that the two phenomena of play and fairness could be biologically related is interesting in its own right, and scientist Marc Bekoff has found it impossible to separate the two in his many studies on wolves.)

In this model, among juveniles, playing is not so much about establishing who may be stronger or dominant, although that could happen as well. Instead, playing is about learning how to be social in a social species. As young lions/bears/wolves/zebras play, both with their peers and with adults, their behaviors are refined through coercion, punishment, and reward. Punishment takes the form of ignoring or temporary isolation of some form. An unruly animal will be shunned by the pack or herd, which is felt quite negatively in a social species, particularly by youngsters. This is the equivalent of the dunce cap, time-out, or being sent to one's room.

Occasionally, adult wolves use mild corporal punishment toward youngsters when an offense is particularly unruly or if urgent correction is needed, such as when they might be hurting another juvenile.[11] In other words, they spank. Harsh punishment is important with young animals because injuries from play can occur. However, such injuries are rare because the adults are constantly supervising, ready to intervene as needed. As they grow older, young mammals learn the rules of proper play and begin to monitor each other. Along the way, they become integrated as law-abiding citizens into the social order of the herd.

This type of training is crucial for social species—animals that live in packs or herds with an established hierarchy and interactive behaviors. (Schools of fish are not generally thought to meet this criterion. We are talking mostly about birds and mammals here—the social-cooperative species.) This is why some animals that are rescued from the wild as youths or born in captivity cannot be returned to the wild later. If they are raised without interaction with their own species, they will not be socialized properly according to the rules and norms of their group. This does not necessarily mean that they will be antisocial. It just means that they will have learned by their interactions with their human captors or whatever surrogate raised them, which probably does not prepare them well to be a properly socialized whatever-they-are.

Thus, we can now understand another solid Darwinian selective advantage of a drive to play in mammals. Animals that play will fit better into the social structure of the group. If they do not play, they are at a disadvantage in the hierarchy, which puts them at a disadvantage for reproduction. Further still, playing properly and observing the rules comes with a biological benefit as well, because if you do not do so, you risk being booted from the group altogether. In the ancient lineage of mammals, the collective benefits of herd living were discovered. Through two hundred million years of evolution, group living led to division of labor, sociality, communicative interactions, and dominance hierarchies. It is quite possible that play was very much at the center of all of this sociality as the means through which young mammals were introduced to each other, to the pack, and to the rules and etiquette of social living.

The socialization of young mammals sounds pretty darn similar to the socialization of human children, does it not? As children, we play all the time, we act out, we are constantly seeking entertainment and input, we are drawn to other children and seek to play with them. As adults, we are constantly "correcting" and honing our children's behavior. Punishments usually involve removal from the social unit—being sent to one's room, being pulled from the game or the playground, and so on. Even in school, kids are sent to stand in the corner or out in the hallway—they are temporarily shunned from the pack! This is how children learn how to behave and how not to behave. It is really not that different from what wolves do.

We also insist on our children playing with other children, especially if they do not have siblings. We call it "socializing," and we are eager to set up various playdates. What we are really doing is subjecting our children to social behavioral training. Common sense tells us—and child behaviorists agree—that it is good for our children's social development to play with other kids, to join the soccer team, even to go away to summer camp. And what happens in these various recreational activities? Children learn to work together with others, follow the rules, recognize and obey authority, and accept punishment for infractions. They also learn the value of the division of labor and cooperative task accomplishment. All while having fun. Think about the first day of summer camp or the first day of soccer practice for very young kids. What is the first thing that is covered? The rules!

Importantly, the lessons learned in Cub Scouts, band camp, and football practice are generalizable to the rest of life. This is the way that our species enjoins the natural instinct to play to the learning of important life lessons. The lessons include how to recognize the pack leader, how to make your own attempt at becoming the alpha, what the rules are and what the punishments are for breaking them, how to make friends and allies, how to avoid or defeat enemies, how to deal with separation from family and loved ones, and how to become emotionally and materially self-sufficient. If you think about it, summer camp is more like survival camp. The lessons learned could rightly be described as savage.

PLAY AS PRACTICE

The summer camp analogy demonstrates the theory that play is useful for teaching social skills and establishing social hierarchies, but there is an even older theory of play called play as practice. These two theories overlap somewhat, and it is important to remember that they are not mutually exclusive. Also, it is time that we look at some species other than wolves and dogs—we have learned plenty from them already.

How about cats? They are not nearly as social as canids (dogs and their relatives). This is quite obvious to any cat owner. Try punishing a cat by ignoring him, and see how far that gets you. It is more like a reward for him. Housecats do not really have a system where they respect "alphas," and even if they did, good luck trying to convince them that the alpha is you. It is extremely difficult to use rewards to teach them tricks, and any rules that they will obey are based on their own wishes more than anything else. They will poop in the litter box only because you have simulated the natural environment that they prefer. (Cats evolved as desert animals, and they bury their poop to avoid attracting predators.) The question is, "Do cats play?"

I am suddenly reminded of an adorable YouTube video: a tiny little lion cub chasing and playing with a grasshopper. Anybody with a housecat has seen this same thing. Kittens are notorious players. They will play with any little thing that they can find: scraps of paper, small pieces of plastic, dust bunnies, and, of course their absolute favorite, bugs. They bat them around, sneak up on them, and pounce. Why do they do this? They are "pretending" to hunt. All kittens do it, from Bengal tigers to domestic shorthairs.

The phrase that some scientists use to explain this behavior is "play as practice."[12] The reasoning here is that play behaviors employed by young animals serve as a warm-up to things they will have to do as adults. Cats are predators, by nature and by instinct. They simply know how to do it. However, that does not mean that they will be successful on their first try. Hunting takes patience and skill. As any human who game hunts will tell you, far more hunts end in failure than success, and it is no different for cats. Learning to hunt involves a lot of trial and error, and a good hunter learns something with each failed attempt. In other words, hunting takes practice. Thus, for predatory animals, the playing they do as children prepares them for the hunting they will do as adults.

If you want to see an even cuter example of animals "warming up" for adult life, visit YouTube and search for "ducks" and "water slide." You will find many videos taken from state fairs around the country in which water slides have been erected so that baby ducklings can play on them. These ducks will quickly run up the ramp to reach the top of the slide and then thrust themselves, slipping and squirming, down the slide, plunging into the pool of water below. Once they have done this, they will swim quickly to the side of the pool and repeat the process. They will do this over and over for hours on end until they collapse from exhaustion.

You do not need to be an ornithologist to see that these young birds are having a blast. They enjoy playing in the water and navigating the water slide. And why? I propose that their duck brains are wired to enjoy various forms of water play, as a means to build swimming skills in a nonthreatening environment. If there were actual danger, the ducklings would be stressed and would not play. Since they sense that it is a safe setting, they are free to simply enjoy the water slide.

I urge you to watch the videos. The ducklings display qualities reminiscent of human children on a water slide—namely, they experience an instant of trepidation when they are at the top of the slide. For a moment, they are not sure if they want to slide down. They look a little scared, but they still want to go down. After a second or two, they summon the courage and take the plunge, inching themselves slowly out on the slide. Anyone who has taken a young child to a water park will immediately recognize this nervous exhilaration.

It turns out that many primate species also demonstrate play as practice, but they do not play-hunt like cats do. This is presumably because most primates are not predator-hunters; they eat mostly plants and bugs. However, most primates are communal animals, living in a complex social hierarchy that must be maintained with behavioral patterns that could almost be described as ritual.

Accordingly, the play that primate youngsters engage in is practice for the behaviors that are used in adult interactions, including mating, child-rearing, and dispute resolution.[13] It has been reported that Vervet monkeys even engage in "play mothering."[14] Juvenile monkeys will engage in mother-like care of infant monkeys in a sort of stylized play. This is not a simple case of contributing to the good of the troop through babysitting, nor is it equivalent to an older sibling helping Mom with the younger kids. The play that these young monkeys do does not help anyone and probably even subjects the infants to some risk. It certainly does not help the mother, who must watch vigilantly. She is not freed up to hunt or even rest. The play-mothering seems to benefit only the playing child, satisfying his or her need to play-act adult life.

The play of human children is not so different from what our primate cousins do. When we were little, we played house, we had tea parties, we pretended to drive trucks or be police officers. The naughty ones among us played—ahem—doctor. I suspect the play-mothering of Vervet monkeys described in the previous paragraph was familiar to some parents.

Toy makers know this all too well. A quick stroll down the toy aisle at your local department store will yield all kinds of big, colorful, exaggerated plastic versions of implements that adults use in their jobs. From Easy-Bake ovens to motorized miniature cars to Johnny's first tool set, mimicry of adult behaviors is a pretty big part of children's play. What looks and feels like playing to us is more than that—it is pretending to be a grown-up and doing grown-up things. What else could this be called except play as practice? How is this any different from the juvenile Vervet monkey pretending to mother a baby or the lion cub hunting the grasshopper?

Is it really so hard to believe that the desire to play-act adult activities is a hardwired genetic drive in human children? It certainly is in other animal species. No one teaches wolves, dogs, cats, or monkeys how to play. They just do it. In the United States, most domestic cats are separated from their mothers and siblings long before they can learn anything from them, but yet, they are playful just the same. It is a natural drive in them, so why is it not a natural drive in us? This makes the evolutionary advantage of play more clear. It is plausible that the playing that we do as children helps to prepare us to be adults—or, at least, that it once did.

Obviously, the reason that children want to play all the time is because it is fun. But why is playing house fun? If you think about it as an adult, it seems like a silly thing to do, but children find it fun. One explanation is that, as children, we are hardwired to enjoy pretending to be adults. The biological benefit of this becomes clearer if we assume that this playfulness actually makes us better at doing adult things. With humans, this is a tricky question. The jury is still out regarding whether or not all that "house" that I played has actually made me better at doing it in real life. But with animals, play-hunting is the first step on the road toward the hunting of real prey.

For full disclosure, the play-as-practice hypothesis has fallen out of favor with some scientists for lack of hard evidence that it truly makes animals better at performing tasks later in life. The reports are conflicting. The biggest problem with attempting to study this in animals is the question of how to deprive them of play without also depriving them of social interactions and other things that may also be important. It is difficult to isolate play behaviors from other locomotor activities and social interactions.

In my opinion, that is the point: play promotes those things by linking them up with the enjoyment-feedback response. It is not so much that the Vervet monkeys truly learn the skills of parenting while engaging in play-mothering any more than human children do. Instead, they gain an outlet to their instincts for parenting and nurture their social interactions at the same time. Nevertheless, some play experts are more skeptical, and that is understandable, given how intertwined all of these behaviors are.

Play as practice is just one working explanation that scientists have developed to explain how and why animals play. There are others as well. The fact that human adults still play argues that there must be additional benefits for play besides play as practice. However, it is true that adults definitely do not play as much as children do, and they play differently. For example, adult play is less about make-believe and more about competition. Sports, chess, poker, and even board games are competitive and skill based. It would be inaccurate to say that adult play has nothing to do with fantasy—just much less so. The fact that the appetite for play wanes during the transition from childhood to adulthood—and has less to do with make-believe—does fit the play-as-practice model very well. Nevertheless, what other explanations for play are there that might apply better to grown-ups?

This is a good time to explain that, when science searches for answers to the mysteries of nature, multiple possible explanations often emerge to explain a natural phenomenon. These differing explanations do not necessarily compete with each other. If there are two possible explanations, both with experimental support, they are not necessarily vying to be the one true explanation. In the case of our study of human play, there might be multiple overlapping explanations, each for different kinds of play or in different contexts and species. The common scientific way to say this is that the multiple theories explaining animal play are not mutually exclusive. They each might be correct for certain play in certain animals. For example, the theory of play as practice works well to explain why human children like to play house, but it may not do as well in explaining why older humans like to play mental games such as chess or physical games such as basketball. There might be other explanations for different kinds of play, and that does not weaken the play-as-practice theory.

PLAY TO HELP ESTABLISH MOTOR COORDINATION

Children and adults both engage in physical play. It is possible that the key to understanding the phenomenon of physical play in humans and other animals is the notion that physical activity promotes brain development, particularly in the areas responsible for skeletal-muscle movements and coordination. In fact, this is a long-known phenomenon and is a key part of childhood development, especially in humans, who are born much less developed than infants of many other species.

Just about the only physical things that babies can do are cry, suck, grasp, and make faces. Incidentally, these are all pretty important for the survival of the babies. Crying is how they communicate their needs, suckling is how they draw milk from the breast or bottle, and making faces is an important part of nonverbal communication. Anyway, besides these few things, babies are born pretty helpless, and their ability to execute complex physical tasks comes slowly over time.

How do humans grow from helpless, uncoordinated babies into suave and elegant ballet dancers? Through practice, of course. As we learn to perform a physical task—whether shooting a basketball or learning a new piece on the piano—we perfect those movements through repetition, and our brains begin to execute them without much conscious effort at all. This is sometimes called muscle memory, but it all happens in the brain. The effect is most dramatically seen when you watch how gracefully and effortlessly a trained basketball player shoots the ball. Compare this with someone who, despite other skills he may have, has never shot a basketball. He will have no form or grace and will clumsily toss the ball in the general direction of the hoop, almost certainly accomplishing the most hilarious of basketball moments—the air ball.

A phenomenon similar to athletic training is at work when infants begin to make their first movements. The connections in the brain that coordinate muscle movements have not been fully developed, and they will not properly develop without many of attempts. Babies' early efforts at controlled movements are rather pitiable—arms flailing about, legs kicking pointlessly, heads flopping around as if they were only loosely connected to the body. Try this with a newborn one time—if she is facing forward and you are off to one side, clap your hands and draw her attention. Watch how long it takes her to accomplish the simple task of turning her head toward you. It would be a really sad affair if it were not also so cute.

But babies do get better. They improve in all their various movements over time as the brain learns from each effort. The motor impulses from the brain to the muscles are fine-tuned using the sensory information going in the reverse direction. This part is unconscious, but it is part of the incredible beauty and complexity of the brain. As you flex your biceps in order to make some intentional movement of your arm, the sensory nerves up and down your arm are feeding information back to your brain, in real time, regarding how the movement is proceeding. Unconsciously, your nervous system monitors all your movements. That is how we gradually get better at physical tasks. This input-output training is essential for the brain to learn how to send the right commands to our muscles that will result in smooth, coordinated movements.

What does this have to do with play? The drive that kids have toward physical play might be nature's way of driving us to be as physically active as possible. The more physically active we are, the better we get at controlling and coordinating our movements. Much of this learning takes place in the cerebellum, a structure that looks like a large walnut located in the very back and bottom of the brain case. Movements actually begin elsewhere, in a region called the motor cortex right on the top of our brains, but they are refined and coordinated in the cerebellum, and there is no other way to achieve this refinement than through lots and lots of repetition.

When toddlers first become able to pull themselves up and stand, they usually spend a lot of time bouncing and dancing, especially when they hear music. This continues pretty much all through childhood—kids are in constant motion. It seems like they never just sit still, as any parent or kindergarten teacher will tell you. This is probably not just a weird quirk. It is likely that we are programmed to incessantly move about in order to ensure the constant refinement of our physical coordination. More movement equals better movement; a physically active childhood yields a more agile and well-balanced adult.

It is well known that the brains of youngsters are much more plastic than those of adults and also more susceptible to the input-output training that develops key brain areas. The acquisition of language is a good example of the plasticity of the young brain versus the inflexibility of the older brain. Childhood physical playfulness may well be nature's way of ensuring that key parts of the brain are properly exercised and developed while the brain is still growing.

If this hypothesis of the benefit of play is true for humans, it should be true for other animals as well. For this, I turn back to the example of cats. Everyone knows how agile cats are. They can jump and scurry and sneak with extraordinary precision and grace—and do so silently. It is truly impressive and awe-inspiring to behold the dexterity of cats.

Anyone who has raised a kitten, however, will know that cats are not born this way. I will never forget watching my first kitten, Sofia, as she was exploring her new home. She would walk around slowly and deliberately. When she would jump up on the coffee table, she did so inelegantly and almost never landed on her feet. "Agility" was not the word that came to mind for describing her clunky movements. I remember very clearly one such event when she struggled to hurl herself up on the coffee table. After a rough landing, she picked herself back up and walked awkwardly along the edge of the table for a few steps before tripping over her own feet and tumbling headfirst back to the ground.

It was a pathetic scene but one that I would never see repeated. Each day of practice served as training for the connections between her muscles, her sensory systems, and her brain. The feedback cycle of inputs and outputs honed these connections until she gradually attained the fine-tuned motor control we all associate with cats. Within a few months, she was jumping from the floor to the kitchen counter, with or without a running start, landing lightly and easily. There is no way that she would have attained that poise and grace had she not worked at it so much as a kitten. Therein is the value of play—training the brain to work with the muscles smoothly and seamlessly.

Physical play is not limited to humans and cats, of course. Mice seem to really enjoy running wheels, and we can find no direct benefit of their doing this. It was previously thought that this was a neurotic or stereotypic behavior, a side effect of captivity, but a recent report showed that mice in the wild were drawn to the running wheels with no reward or incentive and ran on them in sessions that were of similar duration to those of captive mice.[15] Why do mice spend their valuable time running in these silly little wheels? For the same reasons that humans will jump on pogo sticks and play with Hula-hoops: It's fun!

In order to fully appreciate the evolutionary benefit of play as a means to obtaining physical coordination and dexterity, it is important to remember that most species rely on physical agility for their very survival. The same was true for early humans before the invention of agriculture and subsequent civilization. Anatomically modern humans spent about two hundred thousand years living off the land. We had to not only outsmart but also outrun and outfight both our prey and our predators. We were not squarely at the top of the food chain as we are now, and the world was a savage place. The difference between life and death often hinged on how well you moved your arms and your legs. During all that time, the evolutionary drive to play while we were young certainly saved our lives by helping us grow into quick and agile adults.

Two hundred thousand years sounds like a long time, but it is nothing compared with the eighty million years that primates have been evolving or the two hundred million years that mammals have been. All through those eons, physical strength, speed, and agility were vital to the survival of all of our ancestor species and still are for most of our fellow animals. Thus, the drive to acquire impressive physical prowess through playing is ancient and strong. That need and that drive is not limited to childhood, though that is when it is most vital. The persistence of physical play into adulthood is how our ancestors kept their neural connections in top shape. Until recently, couch potatoes would have been eaten by lions.

Even though civilization has recently removed the survival necessity of being in top physical form, the biological drive to play will not necessarily disappear. Why would it? Just because something is no longer necessary or useful does not mean it will simply go away overnight. In fact, a useless behavior might hang around in a species indefinitely unless it is actively selected against. What that means is that unless pointless play starts causing people to die (and thus, not leave offspring), the drive to do it likely will remain. Richard Dawkins once explained it like this: sexual lust derives, at least in part, from the biological drive to reproduce. But that does not mean that lust goes away when we use birth control. Our bodies do not know that safe sex has little chance of producing offspring. Similarly, our bodies do not know that most of us do not really need physical agility in order to survive.

PLAY FOR BUILDING SOCIAL BONDS AND TRUST

A third explanation for the benefit of play is in the development of bonds and trust in social species. Relationships that develop through childhood play will last and help form a closely knit community. If this is true, this explanation for play should function in social animals that live in packs, herds, pods, flocks, or gaggles. Indeed, it has been observed that juvenile primates that play together have been shown to be far less prone to violent conflict as adults, regardless of relatedness.[16] Cubs in a pride of lions are not always siblings or half-siblings and yet, as they play together, they form relationships that will last a lifetime.

Is this true for humans as well? Think about your oldest friends. The bonds that we form playing with other children can last a lifetime and often trump conflicts or disagreements we may have. I can think of several friends with whom I "go way back," and no matter the different turns our lives have taken, the geographic distance, or changes in values or worldviews, I know that we will always be friends and share a special bond for having grown up together. When we spend time together, though many years have passed, we can slip right back into our friendship as we talk for hours about the good ol' days. In addition, these same friends can get away with saying or doing things that I would never tolerate from a stranger or even a "newer" friend. Our bond can survive political disagreements that would normally infuriate me to the point of ending a friendship—simply because of my attachment to our shared history. After all, we played hide-and-seek together when we were six years old.

This sort of social imprinting has some other effects that are also interesting. While these have little to do with play, I want to mention them because they underscore the point that experiences we have in childhood imprint us in a way that affects our adult relationships and behaviors. I know this is not exactly an earth-shattering revelation, but I am not invoking Freud or any deep psychoanalysis. I am talking about more primal human behaviors that have more explicit connections with those of animals.

One of these primal behaviors is mate choice. It is a well-established principle that many higher mammals and birds will actively avoid mating with individuals that they spent a great deal of time with as juveniles as a means to avoid inbreeding.[17] For example, in chimpanzees, our closest relatives, females begin to actively avoid the males that they had previously associated with once they reach sexual maturity.[18] Why would natural selection have favored this behavior? Well, for a young female chimpanzee, all or nearly all of the males that she will associate with during her childhood will be close relatives, most of them brothers or half-brothers. When it is time to start mating, her offspring will fare better if their daddies are outside of the family.

Is this true for humans as well? The incest taboo has some biological basis and is often extended not just to blood relatives but to people that we grew up closely with. We all have people in our lives that, all things being equal, we should be attracted to. Maybe people even ask if we have ever been in a relationship with those people. Because we grew up closely with them, our reaction is something like, "Ew, no. She's like my sister!"

The similarity to the inbreeding avoidance behaviors of our chimpanzee cousins suggests something deeper than a cultural taboo. Studies in humans, most famously the Jewish kibbutz, have attempted to demonstrate what is called the Westermarck effect, named after the famous Finnish sociologist Edvard Westermarck. Often called the first socio-biologist, Westermarck held that humans have a reduced sexual attraction toward individuals with whom they were in close proximity during formative years.

In addition to forming specific bonds, juvenile playing has also been shown to reduce general aggressiveness in rats.[19] In this experiment, scientists raised some rats in an environment in which they were sedated during their social interactions, so they could not play with the other rats, and compared this with rats who could play as normal (but also experienced some sedation, to control for possible side effects of sedation). The play-deprived rats grew up to be easily frightened, defensive, and agitated adults. However, the researchers found that if they provided just one hour of play per day, rats would grow up to be properly socialized with other rats as adults. They also grow up to be smarter, but more on that later.

Playing as a means to establish bonds and reduce aggression is not just for youngsters, either. It is well known that both playing and grooming are used by chimpanzees and other apes of all ages to establish social cohesion and bonding. In fact, a population of chimpanzees in a French zoo is known to engage in pre-feeding rituals that include playing.[20] Why? To reduce the tension and competition that comes with mealtime. Both in the wild and in captivity, meals can be a stressful and competitive event for chimpanzee troops, and fights over food often break out. Thus, playing right before mealtime seems to be an effective strategy to pacify the troop and allow a more harmonious dining experience. If you think that is weird, bonobos sometimes engage in group sex before mealtime, as discussed in chapter 4.

We humans do the same thing. Well, not exactly the same, but similar. While mealtime is not as conflict-prone for us, humans are still very much a competitive species. Competitions can lead to conflict, and conflict is counterproductive. Since the dawn of civilization, within-group interpersonal competition has frequently reared its counterproductive head in the workplace. To mitigate this, it seems that employers have taken a cue from those French chimpanzees. Anyone who has worked in a corporate or academic setting can tell you about periodic "retreats," team-building exercises, and the like. What do the employees do at these events? They play together. The stated purpose of these activities is to promote social bonding and cooperation, to reduce conflict and tension, and to facilitate teamwork and personal relationships. Just like chimpanzees.

Playful competitions for reducing tension and aggression can even be scaled up to the global level. I am talking about the Olympic Games, of course. The Olympian spirit is explicitly oriented toward reducing global conflict through sport. It is up for debate whether or not it actually works, but there is a solid biological basis for how it might. Playing builds bonds of trust, in children and adults, and the mechanism could be that the release of all those joyful neurotransmitters in our pleasure center creates an opportunity for intimacy and understanding.

PLAY FOR MANAGING STRESS

We all recognize that some forms of play and recreation are good for relaxing. We use our favorite pastimes as welcome escapes from the pressures of our daily lives. They relieve our stress, at least for a little while. Is it possible that our perceptions about recreation relieving stress are not just superficial? Could it be more than just a psychological effect? Is there a medical benefit to getting some quality playtime now and then? Is this medical benefit strong enough to have resulted in survival advantage? If so, there is a good case that we have evolved to enjoy recreation at least in part because of its therapeutic value.

When it comes to relieving stress, not all types of play are the same. For relaxing, we are usually talking about things like reading, watching a funny or interesting movie, gardening, tending to a collection of some sort, and other hobby-type recreations. We all have a few activities that we do purely because they relax us. Most physicians agree that stress is bad for your health and that the periodic relief of stress through relaxation has medical benefit. We all know this intuitively, but there is good science behind it, too.

A great deal of evidence has indeed linked the levels of circulating stress hormones (mostly cortisol, but also epinephrine and aldosterone) to general sickliness. Cortisol is known to directly inhibit cells of the immune system, so it is no surprise that chronic stress leads to colds, the flu, and other minor respiratory and digestive infections. Even minor cuts and scrapes heal more slowly in a stressed person. The general immunosuppressive effect of stress works through many mechanisms, including reduced antibody production, reactivation of latent viral infections, and the general inhibition of white blood cells, which are our primary infection fighters.[21] Scientists have actually observed stress hormones restrain white blood cells in a laboratory setting.

Since stress hormones make you sick, and relaxation reduces stress hormones, it stands to reason that putting time aside to enjoy relaxing activities would be good for your health. Science has confirmed the old wives' tale that working long hours will weaken your immune system and make you more susceptible to seasonal colds and flu.[22] Indeed, the sickest I have ever been was when I got mononucleosis at the end of my first year in college. Lest anyone wonder if it was something else that led to my getting the "kissing disease," in my defense, I had been up late studying for finals every night for a week beforehand. The stress, combined with the lack of sleep, made me a prime host for the Epstein-Barr virus that likely caused the mono. (No matter how I was exposed to it.)

However, not all forms of play and recreation can be described as relaxing. Think about a paintball competition, an intense game of basketball, most video games, a high-stakes poker game, even a round of golf. Far from relieving stress, these activities actually induce a great deal of stress. Are these forms of play actually bad for our health? If so, why would we have evolved to want to do them? These forms of competitive play are probably the most similar to animals that wrestle and joust for fun. It almost looks like fighting, but it is not. It is competitive play. But if it causes stress, why would this have evolved?

It turns out that even stressful forms of play are actually relaxing. The secret to this apparent paradox is in the difference between short- and long-term stress. Short-term stress, sometimes referred to as sympathetic neural stimulation or the fight-or-flight response, involves a different set of hormones than long-term stress. While long-term stress releases cortisol and aldosterone (among others), short-term stress releases norepinephrine and epinephrine (adrenaline). The result is that any inhibition of the immune system during acute stress is only temporary. The body returns to normal very quickly. In addition, your brain and body seem to know the difference between real life-or-death stress and the playful form of stress experienced during a friendly competition. This is why this form of play does not violate the last criterion of play, which requires animals to be in a "relaxed field" and free from (real) stress.

Even more important, following the fight-or-flight stress of a game, our long-term stress hormones are actually reduced.[23] It is as if the burst of "safe stress" is opening some sort of stress valve in our bodies, lowering our baseline level of stress over a longer period of time. We even sometimes refer to this as "blowing off steam." Similar to a cardiac defibrillator, by delivering an enormous depolarizing shock, a normal resting heart rhythm can be reestablished. I know I am not the only one that feels much more relaxed after a long run or a round of golf than I did beforehand. The point here is that both relaxing and intense forms of recreation can reduce the hormones involved in long-term stress and thus both are good for our health.

This health benefit of play is almost certainly not unique to humans. First of all, the effect of stress hormones on the immune system is well documented in all sorts of laboratory animals. Second, in addition to the effects on the immune system, physical play and exercise are well known to promote good cardiovascular health in both humans and animals. Plenty of research shows that getting enough rest and minimizing stress is a recipe for longevity in both laboratory and zoo animals. This is somewhat difficult to dissect. Does the health benefit come from the exercise itself by way of increased physical fitness? Or is the reduction in stress hormones responsible for most of the health benefit?

In a certain sense, it does not matter. Whether it is the cardiovascular improvements or the reduced stress—or both—the point is that it enhances your survival. And if it enhances your survival, natural selection could act to promote these behaviors through a natural drive to play or exercise. I suppose that we may never really know whether we have a natural drive to exercise as adults or if that drive comes purely from conscious health or beauty concerns. The bottom line is that play is good for us—and always has been.

There is one more aspect to the stress theory of play. Some scientists have theorized that, in children, the stressful aspects of competition and intense play help condition us to manage stress later in life. By exposing us to this safe form of stress, "tense play" helps prepare us to deal with real stress and keep our cool while doing it.[24] Real-life stress can involve matters of life and death, which was especially true in our prehistoric forebears. If acute stress made someone freeze or panic, that could have meant his demise. Similarly, if chronic stress causes someone to gradually go to pieces and mismanage her life, she is in deep trouble.

In contrast, the most successful among us are those who rise to meet challenges and never shy away from stress. In fact, many high-performing people report that they are at their best under conditions of stress, such as deadlines, contests, and intensity. It is conceivable that this is a skill gained while playing games. By easing into a life of stress through playful "safe stress," we may condition our bodies to cope with real stress and even thrive under it.

Once again, it is the lab rat that provides strong evidence for this possible benefit of play. While not specifically addressing play, researchers have found that rats that have been periodically stressed when they were young cope much better with stressful situations as adults. When placed in novel situations, the stress-accustomed rats "froze" less, explored more, and were generally less fearful than rats that had not been forced to deal with stress when they were younger. Interestingly, these "stress-primed" rats had a less-pronounced surge of cortisol when they were placed in stressful situations later on.[25] Thus, most scientists agree that mild stress can actually be good for us, if engaged safely and with moderation. That is exactly what competitive play does: it gives us a safe outlet for learning to manage stress.

PLAY FOR DEVELOPING CERTAIN COGNITIVE AND CREATIVE SKILLS

Does playing make us smarter? The answer appears to be yes. As I will discuss shortly, playing as children helps develop certain abilities and overall intelligence, while playing as adults helps to "refresh" some skills that may have become exhausted and gives a burst of creativity. In fact, there is a pretty good correlation between the size of the forebrain, where higher cognitive functions are housed, and the tendency toward playfulness among animal species. In other words, the bigger the brain of the animal (relative to body size), the more the animal will play. This begs the question of cause and effect: do big brains make us play, or does play give us big brains? As usual in biology, the answer is likely a heavily nuanced "both."

Beginning with animals, as I alluded to before, laboratory rats that are denied the opportunity to play suffer deficits in the development of their brains.[26] Other studies have shown that rats with certain damage in their prefrontal cortex will not be driven to play as much, and when they do, they do not perform as well in competitive play. In turn, these rats have developmental defects in skills that would normally have been honed through play.[27] These are not just a few isolated studies that can be interpreted any number of ways. It turns out that the importance of play to the cognitive and neural development of rats has been studied many times and in many ways.[28] The summary of all of this research is that play is a key aspect for the proper development of rats, not just socially but cognitively as well. Playing is required for full and normal development of the brain and overall intelligence in rats.

Similar observations have been made in other animals as well. Play helps animals develop spatial mapping skills—the ability to visualize objects and places and their relative position to other objects and places. It helps them solve problems such as mazes and perform other learned skills. Playful animals will be better at manipulating objects successfully as tools. The list goes on and on. There is every reason to believe that playful behaviors are important for the normal brain development of all animals.[29] The best animal in which to demonstrate this principle is the one with the most cognitive development to worry about in the first place: humans.

When it comes to humans, there is a staggering amount of evidence that playing is important for cognitive development. Many books have been written on the subject, and the educational psychology of play is a vibrant academic subdiscipline unto itself. For constraints of space, I will attempt only to briefly summarize the major points of what we know about the value of play in human children. Experts in play generally talk about seven modalities (types) of play, each with its own benefits:

Attunement refers to play between infants and their parents/caregivers. This type of play involves babies just looking and being entertained, but it is crucial for emotional development because at this stage emotional states are almost purely "contagious" for the babies (see chapter 4). When you hold an infant and smile and express joy, the infant often "catches" the joy and shares in it. Similarly, fear and stress can be conferred upon the baby as well. In a sense, young childhood is when we develop our most basic emotions. Neglected children almost always suffer emotional difficulties, and this is part of the reason why.[30]

Body play is the physical play that we have already discussed for its role in developing coordination and locomotor precision. It also helps with visualizing spatial relationships and with understanding how our body relates to the world around us, which is important for learning to avoid physical dangers and for respecting natural boundaries.

Object play is playing with toys and other physical things, mostly with our hands but also with other body parts. This, too, helps with understanding spatial orientation and visualization in three dimensions. It also develops reflexes and fine motor control.

Social play is that which involves others and is useful for learning social structures, etiquette, and rules.

Imaginative play is among the most intriguing forms of play. It was touched on earlier in the play-as-practice section, but psychologists tell us that it is much more than that. It is believed that imaginative play is key to developing the broad range of cognitive abilities in humans that involve abstract concepts, independent thought, and creativity.[31] Children use make-believe to explore and interact with the social roles that they are gradually learning in their environment. These include gender, professions, and various activities that they see around them. Children will vociferously engage in imaginative play with few or no props needed.

Through make-believe, children develop their own imaginations and learn the ability to perceive things that do not actually exist, both figuratively and literally. In the figurative sense, they can imagine scenarios that do not exist, like being a mother, a professional athlete, or a teacher. By literally, I mean that they practice the skill of mental imagery where they close their eyes and conjure an image. When we do that, we perceive something visually, even though we do not actually see it. Did play lead to imagination, or did imagination lead to play?

Narrative play is the process of storytelling, both in reception and delivery. This, too, exercises the imagination, but it also promotes the development of language skills. Language development will be discussed in chapter 10.

And finally, *transformational play* occurs when we take on a new identity and then solve problems and challenges as that new identity. This often involves a different set of rules and surroundings from our actual environment and thus requires creative thinking and "transcendence" from some of the limitations of our actual world. In a sense, we repeat the process of our earlier "probing" of the limits and rules of our actual world and apply that same process of discovery in an imaginary setting that we do not totally control (as in purely imaginative play). Video games are good examples of this type of play. Sorry, parents, many studies have shown that video games can promote development of certain cognitive skills and even raise children's intelligence in some contexts.[32] Of course, these benefits should be balanced with the benefits from physical and social play in order to shape a well-rounded and well-adjusted child.

The bottom line here is that there are many benefits to the various types of play when it comes to developing the full potential of the human brain. The sensory-motor stimulation, the imagination, the socialization—all of this is good for human development.

What about adults? Is playing good for adults as well? The adult brain is much more hardwired and less able to pick up new cognitive abilities, so it is at least possible that play is not as important for adults. Indeed, the drive to play is much weaker in adults than in children, but it does exist. We adults play, too. In addition to the stress relief and cardiovascular health benefits, are there also cognitive benefits for adults?

Play has been found to be a very effective way to enhance learning in students of all ages.[33] Many studies have shown that physical exercise is good for maintaining cognitive function in senior citizens, but what is often overlooked is that the exercises that are most effective for this are games and team sports, which combine body play with social play and object play. One study even showed how the Nintendo Wii improved the well-being of the residents of a retirement home.[34] Finally, sessions of play are also associated with bursts of creativity and successful problem solving.

* * *

While researchers are divided as to the origin of play, there is no doubting the biological benefit. Squirrels that play more are better coordinated and better parents.[35] Play has been shown to be beneficial for rats, brown bears,[36] and wild horses.[37] Play has been shown, in humans and animals, to promote intelligence and, specifically in humans, original and creative thinking. Stuart Brown, a physician and one of the world's leading experts on play, put it this way: "The opposite of play is not work—the opposite of play is depression."[38]

Even in the time of Plato, it was known that humans are hardwired for play. This propensity gradually, but not completely, fades with maturity. Because there are multiple benefits of play, it is no surprise that the behavior is widespread in nature. However, as with everything else in life, there is a natural tension between the advantages of play and the disadvantages of overdoing it. Play may be good for a young lion cub, but only up to a point. If she gets so distracted so as to become unaware of her surroundings, a hyena might make an easy meal of her. And while all work and no play may make a bonobo dull, all play and no work will get him cut out of the food-sharing hierarchy or shunned altogether.

These opposing forces of play and more "serious" endeavors have been fighting it out for millions of years. This, too, sounds familiar. In a sort of reenactment of evolution, many of us will spend our adult lives trying to strike the perfect balance between work and play. Many educational psychologists believe that our children would learn better if we employed more play in our schools—playing to learn. Similarly, in today's very corporate culture, most experts of play agree that we have suppressed fun and games too much among adults. Maybe that is why we have not seen as much of a decline in chronic illnesses as we might expect, given our advances in preventive and curative medicine. In other words, "When we stop playing, we start dying."[39]

FURTHER READING

Brown, Stuart L. *Play: How It Shapes the Brain, Opens the Imagination, and Invigorates the Soul.* New York: Penguin, 2009.

Burghardt, Gordon M. *The Genesis of Animal Play: Testing the Limits.* Cambridge, Mass.: The MIT Press, 2005.

Byers, John A. *Animal Play: Evolutionary, Comparative and Ecological Perspectives.* Cambridge: Cambridge University Press, 1998.

2

ANIMAL SYSTEMS OF JUSTICE

T HE HUMAN FASCINATION with justice is as old as civilization itself. Some of our most cherished and celebrated documents reflect this: the U.S. Constitution and the Universal Declaration of Human Rights are two of the more recent examples. Our tradition of codifying our rules of conduct goes back to the Magna Carta, the Old Testament, and Hammurabi's code. Human society is obsessed with laws and justice. Even when we gather to play games, we first establish the rules. We cannot even *have fun* without also imposing rules.

But what is justice? Is it nothing more than a man-made set of rules that we have chosen to impose on our neighbors and ourselves in order to promote social order? Does justice stem only from civilization? Is a sense of justice a unique feature of very high-level cognitive ability? These questions focus solely on the human application of justice in a formal setting, but is there something deeper than that? Is there a biological basis for the concept of justice?

One thing about justice that is particularly striking is how similar it is in diverse cultures and histories around the world. Too often, we get caught up in the contrasts between cultures, but the similarities far outnumber and outweigh the differences. Of course, those similarities could come from social forces or just common sense about the need for stability, fairness, and harmony in society. However, they could also come from our biology. Could it be that we are programmed to recognize and desire justice?

INTOLERANCE OF INEQUITY

Researchers at Georgia State University led by Sarah Brosnan recently conducted a few simple but telling experiments involving chimpanzees.[1] A group of chimps had been trained, individually, to retrieve and then surrender tokens in exchange for food treats. The price was set: one token bought one piece of food. Sometimes the food was a grape, and sometimes it was a carrot piece of similar size. Like most of us, the chimpanzees preferred the grape and would select it over the carrot if given the choice. However, if presented with only the carrot, it was a prize they would happily eat in exchange for the silly token that was of no use to them otherwise.

Everything went fine in this experiment until two chimpanzees were brought in together and allowed to earn their treats in full sight of each other. If the two chimpanzees repeatedly earned the same reward, be it carrots or grapes, they were happy to continue the task, each largely ignoring the other. However, when one chimpanzee earned a carrot after seeing another chimp earn a grape, more often than not, he did not accept it. Instead, he would refuse the carrot, grow restless, show obvious signs of agitation, and perhaps even throw a tantrum. Before, he just preferred the grape but was happy with a carrot. Seeing another chimp get a better reward for the same task was more than he could bear.

I doubt this scenario will sound at all unfamiliar to any parent with two children. The perceived value of a treat or reward can be strongly influenced by the treats or rewards that others receive. And this human feature is in no way limited to children—adults are just as susceptible to this kind of behavior. But the question is, should we be surprised that it is seen in animals such as chimpanzees?

For whatever reason, the image of a chimpanzee refusing a reward because it is not as good as the reward received by another just seems so *human*. I find this curious, too, because it is not exactly a mature or sensible way to react. It is kind of obnoxious, actually. (Therein is the resemblance, I suspect.)

Brosnan's group provoked the chimpanzees further. In another round of experiments, researchers would first show the chimpanzees grapes, but when they returned with their tokens to purchase them, they were instead given carrot pieces. Again, the chimpanzees very often refused the carrot pieces, regardless of their previous willingness to earn the carrots by performing the same tasks. They recognized the unfairness of the situation and refused to play along.

Something about this experiment seems almost cruel, like when someone teases their dog by saying words like treat and walk but then does not follow through. But that is not what the experiment was—there was still a reward there for the taking. All the chimpanzees had to do was accept it. They refused to do so because the reward did not measure up to the expectation that they had been led to. They thought they had a deal, and then the terms were changed on them. In human terms, this is called the bait and switch, and it turns out that chimpanzees are no more amused by it than we are.

We might be inclined to chalk these two examples up to good old-fashioned greed—the chimpanzees wanted the grape, plain and simple, and knowing it was out there made the carrot look less attractive. However, the researchers also found something that I doubt they expected. Sometimes—not all the time, but sometimes—the chimpanzee that was given a grape (the better reward) would refuse it if he saw that his buddy got only a carrot. He would refuse the better reward because someone else had been cheated. Clearly, this refusal goes beyond greed. The chimpanzees could recognize the inequality going on, and they would not stand for it, even when they were the beneficiaries. Some of these chimpanzees preferred to go hungry and throw a tantrum rather than tolerate the unfair conditions.

If this is not the beginning of a system of justice, then I do not know what is. Equal pay for equal work is what we are talking about here. And it turns out that this principle is not just some constructed ideal or the product of an advanced and progressive society. It is part of the natural psychology of us and our closest relatives.

The chimpanzees were clearly experiencing a conflict. Their own self-interest should have told them to take the reward—any reward—rather than risk getting nothing by throwing a tantrum. Instead, they exercised a different value: fairness. Unlike humans, no one ever sat these monkeys down and taught them the Golden Rule or told them that God was watching them—they just knew it was wrong. Something resembling a sense of justice therefore must be hardwired into the chimpanzee brain.

Biologists refer to this behavioral trait as "intolerance of inequity." If chimps and humans both understand fairness, this means that it likely evolved in some common ancestor of the two. The most recent common ancestor of both chimps and humans (and bonobos) lived around six or seven million years ago. The concept of fair pay was around a long time before the first humans walked the Earth.

The work with chimpanzees described here was not the first of its kind. In fact, Professor Brosnan, while studying under the mentorship of Frans de Waal, had previously performed a similar experiment with a species of New World monkeys called capuchin monkeys. Although the experiments with the capuchins did not probe as deeply as those with the chimps, the researchers found that the monkeys would happily turn in their tokens for cucumber pieces—unless they had previously seen another monkey receive a grape for performing the same task.[2] (You can watch a YouTube video of a capuchin accepting a cucumber, then later rejecting it—even throwing it at the scientist—after seeing another monkey getting a grape. Search for "capuchin monkey fairness experiment.")

That such similar behaviors are observed in two primate lineages that have been geographically and genetically separated for so long points to a common root cause that is at least as old as the evolutionary branching itself—and by root cause I mean genetics, of course. Since New World monkeys have been evolving separately from apes for at least forty million years, we can assume that primate species have understood fairness for at least that long.

Even more recent work with chimpanzees reveals the same sense of fairness but in a different scenario, the ultimatum game (UG). Variations of the UG are commonly used when studying the psychology of people, especially children, and in something called game theory. In its simplest form, the UG goes like this: Person X is given an amount of money but has to propose how to split it between himself and another person, Person Y. Once X has made an offer for the split, Y can choose to accept or reject it. If he rejects, neither party gets anything. Although there are several interesting things to probe about human nature with this experiment, most commonly we focus on Person Y. Once the offer is made, whatever the offer is, Person Y should accept it because to reject it means he gets nothing. In practice, however, most of the time, Person Y rejects the offer if it is less than around 20 percent of the total amount.

Even if the offer is only 10 percent of the total, that is better than nothing. Would the rational choice be to accept whatever offer is given? Well, as we all know, humans are not purely rational. We have other forces at work that influence our decision making, and one of those forces is the intolerance to inequity, as shown by the UG. The UG has been used to study intolerance to inequity in humans for many years, so why not try it with chimpanzees?

That is exactly what Professor Brosnan's group did. They invented a new version of the UG to use with chimpanzees.[3] In this setup, Chimp X was given two tokens to choose from, both of which would earn him some bananas. However, he had to give whichever token he chose to Chimp Y, and then Chimp Y had to submit it to the researchers so that the chimps could get their reward. The catch was that one of the tokens represented an even split—three bananas for each chimp. The other token earned five bananas for Chimp X and only one for Chimp Y. Remember that Chimp X could not turn the token in himself; he needed the help of Chimp Y.

Under this scenario, Chimp X very quickly decided to pursue the "fair" course of action. However, if Chimp X needed no help from Chimp Y and could do whatever he liked, sometimes known as the dictator game, he submitted the token that got him the larger reward, with no regard for the other chimp. In other words, if cooperation was needed, Chimp X knew that he had better pursue the fair option, rather than the unequal option.

Although Chimp X was motivated by self-interest in both scenarios, the difference in behavior is striking because it reveals that he recognized that the two scenarios were different. In other words, he appreciated the difference between equality and inequality. Furthermore, he recognized that if he needed the cooperation of Chimp Y, equality was the best choice. In this way, fairness is a value in chimpanzee social behavior, but only insofar as it aids cooperation. This phenomenon highlights what will become a recurring theme in this book: evolution and natural selection are not forces that favor only selfishness and brutal competition. Sometimes, one's self-interest is actually served through cooperation and fair play, and many nonhuman animals recognize this.

The interesting question is this: What motivated Chimp X to share the bananas equally? What was he afraid would happen?

A doubter of these results might say that this was just a cold calculation on the part of Chimp X. He is betting that if he chooses the "fair" option, thus making a short-term sacrifice, it will lead to better rewards in the future because it will maintain the cooperation of Chimp Y and thus keep the banana train rolling. Maybe it is still selfishness, just a more sophisticated and calculating form of it. My response is that that is what fairness is. It may very well be that justice is simply a more comprehensive and more sophisticated form of self-interest.

We do not insist on rules to be sure that we do not cheat; we insist on rules so that *others* do not cheat. Our sense of fairness expresses itself much more strongly when we are the victims of unfairness than when we are the beneficiaries. I am not suggesting that there is no such thing as a selfless act, although many others have. What I am saying is that at least part of the motive of fairness comes from self-interest. The greatest of all rules in Christianity is called the Golden Rule: do unto others as you would have them do unto you. Even this rule, plainly stated, asks us to reflect on repercussions for ourselves. We are asked to consider how unfairness would affect us before we inflict it on others—the idea being that, if we all live by these rules, we will not be victimized.

What about species other than monkeys? It turns out that we primates do not stand alone with our intolerance of inequity. This behavior is present in our best friends as well. A 2008 study with dogs demonstrated this quite clearly. In that study, when dogs were asked to "shake" (aka "give paw"), if they knew they would get a treat, they would comply with the request every time. No surprise. But if there were no hint of any reward, they would only satisfy the request about twenty times out of thirty. However, if they saw other dogs getting a treat, while they did not, for completing the same assignment, they would complete the task just twelve times out of thirty. In addition, the researchers observed visible agitation, barking, and aggression among the dissatisfied pups.[4] It sucks to not get a treat. It sucks even worse to not get a treat while others do.

Intolerance of inequity is being observed in a diverse and growing list of mammals, which would push its origins back even further to well more than one hundred million years ago.[5] Why would this behavior have evolved in the first place? It is difficult to imagine that throwing a tantrum every time you perceive being slighted would lead to increased survival and reproductive success. It is also hard to imagine any good coming from it at all. What purpose does this intolerance of inequity serve for survival and success?

COOPERATION AND PROSOCIAL BEHAVIOR

Once again, we find ourselves faced with the notion that the rules of natural selection may not be able to explain a behavior. Would survival of the fittest reward the cheaters and punish the righteous? Actually, if we consider the long-term effects of demand-for-justice behavior, a distinct Darwinian advantage emerges: fairness is essential for cooperation, and individuals who cooperate fare better than those who do not, at least in certain contexts and in certain species.

In a later chapter, we will see how birds and mammals have evolved reproductive behaviors built around cooperation, instead of just competition. It turns out that animals, especially mammals, often cooperate in many other ways as well. I am focusing my discussion of fairness and cooperation on mammals because, while cooperation is not unique to mammals, our class of animals has definitely taken it to the highest level. Most mammals are social species living in communities from tens to hundreds of individuals. Even those mammal species that have returned to individualistic and solitary lifestyles at least begin their lives in the social context of nursing and weaning. (The word "mammal" comes from the Latin root *mamma*, which means breast.)

For a social species, cooperation is essential to living together peacefully. The community cannot thrive if individual members try to sneak more than their fair share or get away with not pulling their weight in the herd or pack. In order to enforce discipline and acceptable conduct, there must be a genetic drive to punish those who are out of line.[6] If and when selfish or uncooperative individuals are ostracized, they will likely suffer elimination from the gene pool, as members of social species generally do not fare well on their own. Natural selection would thus select against the selfish through their banishment and subsequent failure to reproduce. However, before a social species can discipline unruly members, it first must be able to recognize injustice when it sees it. This is the intolerance to inequity that researchers have found.

Behaviors that focus at least as much on the good of others as of ourselves are called prosocial. Scientists have classified plain old cooperation separately from truly prosocial behaviors because cooperation can be purely selfish in nature. "If we help each other, I eat. If we do not, I will go hungry. Therefore, I will help you." Prosocial behaviors, on the other hand, come with no direct benefit to the individual, only to others or to the group. These kinds of behaviors were once thought only to be found in humans.

However, chimpanzees have now also been found to exhibit prosocial behaviors. In a recent study, chimpanzees were asked to choose between two tokens. One token got them a small reward, nothing more. The other token earned them the same reward but also allowed a second chimp (unrelated and unfamiliar) to get the reward. The two chimps were separated but could see and hear each other and, importantly, never switched places, so there was no immediate chance of reciprocity. What the researchers found was that the chimpanzees were happy to help out a stranger by selecting the token that led to a reward for both chimps.[7] "Why not? No skin off my back!" This is prosocial behavior. No one is lining up to give them a Nobel prize, but choosing to benefit others with no gain for yourself is the definition of prosocial behavior. It is not, however, altruism. Altruism requires a cost for the generosity, and there was no cost in this situation.

Interestingly, the researchers found that the behavior of the second chimp—the one that benefited from the prosocial generosity—affected the likelihood that the first chimp would choose to help her. (All the chimps were female in this experiment, for simplicity.) If the second chimp called attention to herself by tapping or vocalizing in a gentle, nonthreatening way, the first chimp was even more likely to help. This could be considered asking or begging. However, if the chimp was loud or aggressive, jumped around, or was obnoxious in any way, the first chimp was substantially less likely to help. Ask, and ye shall receive. Demand, and you can go $#@ yourself.

This, too, seems familiar. Imagine the following scenario: you are walking down the street, and someone just in front of you drops a couple dozen pencils on the sidewalk. Maybe you would not stop, but you probably would. This does not cost you anything, so it is prosocial to stop and help. However, what if you were late and in a hurry? In that case, there would be a cost to you, so this would be altruism. Under these conditions, you would be less likely to stop. After all, there are other people who can help, right?

Further still, if the person who dropped the pencils looked at you pleadingly and desperately, you would probably stop, regardless of whether or not you were late. On the other hand, if this person yelled at you and demanded that you stop and help, you would probably walk on by, even if you had all the time in the world, right? This is such a simple scenario, but it reveals a rather deep connection in the prosocial behavioral impulses between humans and chimpanzees.

HANDICAPS AND APOLOGIES

The story does not end with a sense of equality of resources. Other manifestations of fairness are seen in nature as well. Professor Marc Bekoff has published many articles detailing the rules of fairness that govern wolves, particularly when they play. One striking phenomenon is the self-imposition of a "handicap" when a larger and stronger wolf plays with a smaller one. The larger wolf will approach the smaller one in a submissive posture, which gives the former a distinct disadvantage in the ensuing wrestling match.[8]

Why would the larger wolf handicap herself? Well, if you are a big, strong wolf, sometimes the only way to find a playmate is to agree to level the playing field. This, too, hints at a sense of fairness. Both wolves, big and small, recognize the advantage that the larger wolf has. If you were a small wolf, why would you want to start a wrestling match when you know you were going to get your clock cleaned? Nevertheless, as we saw in chapter 1, the instinct to play is hardwired. Both wolves really want to play, but they also recognize the inequality of the pairing. This poses a conundrum. The self-imposition of a handicap is the solution to the conundrum. The larger wolf puts herself at a disadvantage so that the smaller wolf will play along. In fact, the stronger wolf will also allow the smaller wolf to bite her, which she could easily prevent, provided the play bites are not too hard.[9]

Perhaps even more telling is that when young wolves play, if one wolf goes too far and bites hard enough to cause injury, the game is stopped and the offense must be apologized for. Wolves have a way of offering apologies. The offending wolf will approach in a submissive posture, hanging her head low in what is called an apology bow. If an apology is not made after the offense, no wolves will play with the offender again. Given how important these play fights are for the social cohesion and dominance hierarchy of the pack, this is a dire consequence. A wolf with whom no one will play subsequently loses any hope of a dominant position in the pack.[10]

The same system of handicapping during play-wrestling is also seen in coyotes, close relatives of wolves. However, with coyotes, the punishments are even more severe. A single offense of biting too hard can get an individual permanently shunned from the pack.[11] Although coyotes are often solitary as adults, when youngsters are ostracized too early, they suffer a much higher mortality rate, making this punishment quite harsh. You break the rules; you pay the price. Justice, it seems, is not just about fair play. There must also be punishments meted out to the offenders, and forms of social punishment have been documented in a variety of species.[12]

This submissive-apology behavior is also seen in domesticated dogs, the direct descendants of wolves.[13] We see the "I'm sorry" posture any time a dog gets into the trash, has an accident in the house, or, in perhaps the most resonant example, gets too rough with the kids. Importantly, while it could be argued (unsuccessfully, it turns out) that apologizing is a learned behavior in wolves, no such argument could reasonably be made for domesticated dogs. No one sits their dachshund down to teach him how to submissively apologize for an offense. He just knows. There is some sort of "guilty gene" at work, again demonstrating the genetic basis of complex behaviors.

Once again, I am reminded of my family sheepdog, Daisy. We had a large room in the house where Daisy was not allowed to go. The dog-free formal family room was for when we had visitors who did not particularly enjoy a big dog jumping into their laps (like Grandma and Grandpa). Daisy never hid her resentment of this rule and would frequently stand or lay down just on the threshold of the room, as if to test the rules and inch her way across the boundary. If she was ever suffering from a stomach issue or we inadvertently left her too long without a bathroom break and she could not hold her bowels any longer, without fail, she would relieve herself in that room—the dog-free room. Every single accident she ever had inside the house took place—with no one looking—in the one room that she was not allowed to enter.

Did she choose to go in that room in order to get back at us? Was this some sort of revenge for imposing the rule that she hated so much? Or was this her attempt to cover her tracks and fool us into thinking that it was someone else? "Guys! Surely it wasn't me that did it—I'm not even allowed in there, remember?"

Whatever her reasons, there was no hiding her actions. The guilty look gave her away every time. Upon our return, as soon as we opened the door, if she did not run to greet us, we knew something was up. Inevitably, we would find her sitting in some corner of the house, giving her most convincing apology bow. The guilt program deep inside her wolf brain had been activated, and she was desperate to receive the punishment necessary to satisfy that guilt, allow forgiveness, and restore harmony in the social unit. She was just like a wolf, and we were her pack.

Clearly, handicaps and apologies seem to be ingrained features of the social lives of canids, the group that includes wolves, coyotes, domestic dogs, dingoes, and jackals. Investigators have also observed self-handicapping in five different monkey species.[14] Interestingly, scientists studying chimpanzee play have found that the tendency to self-handicap appears spontaneously and early in childhood.[15] Chimpanzees do not learn this behavior, either by instruction-demonstration or through negative consequences following infractions. Although it could be fairly said that positive consequences reward the self-handicappers when they are able to better attract playmates, the point remains that the behavioral instinct to do this is inborn.

Self-handicapping thus seems to be a common, even near-universal feature of physical-competitive forms of play in mammals, including our closest primate relatives. Is this something that we humans do as well? I think it is pretty obvious that we do. When amateur golfers regularly play together, they compete against each other utilizing their respective "handicaps" to level the playing field. Their individual skill levels are normalized and written away in order to establish a new common baseline. This makes the competition fair. Golfers with very different skill levels can play together and compete because they begin from different places based on their handicap level.

In fact, the word "handicap" was originally developed for exactly this purpose, rather than to mean a physical limitation as it came to be used later. There was a bidding and trading game in medieval England called hand in cap that involved an umpire establishing monetary valuation of possessions being bid on by two competitors. Somehow, the practice of a third party establishing baseline weighting between two unequal competitors became known as handicapping, and it was applied to horse races as early as the eighteenth century. Horses with stellar racing records were "handicapped" by having weights added to their saddles. And voilà—the word "handicap" emerged as something used to weaken an otherwise stronger individual.

For some reason, this term has fallen out of favor when referring to individuals with a physical or mental impairment. I suppose it could be because of a false impression that the word handicap means "not as capable." There is also the apocryphal story that the word handicap began as a pejorative for beggars who had their "caps in hand," an urban legend that appears to have been invented out of whole cloth. I feel that if more people understood that the word handicap originated as something applied to overachievers to make things more competitive for the rest of us, maybe it would not have been replaced with words like disabled. I cannot for the life of me see how the word disabled came to be preferred over handicapped, but I digress.

My point here is that the concept of handicapping is very common in human culture, especially gaming. In high school sports, there are different "classes" in which the teams compete so as to ensure that a tiny rural high school is not routinely obliterated by a huge urban powerhouse. In professional sports, the draft, where new players are selected, is heavily tilted to favor the lowest-performing teams from the previous season; boxing is organized by weight class; and so on. The point is that we prefer a fair fight to a lopsided one.

Further, the handicapping of stronger entities is not limited to sports and play. Our system of taxation employs this as well. Every single industrialized nation employs a progressive tax code in which the more income you have, the greater the marginal tax rate you are subject to. (At least, that is how it is intended to work.) Similarly, economically disadvantaged individuals and families are given government support for food, housing, and medical care; a baseline of education is provided to all children at no cost; and so on.

Our justice system also incorporates features of handicapping. For example, in the United States and most other Western countries, there are laws against price fixing that are designed to ensure a fair contest among competitors and prevent coalitions and scheming. There are also related laws against monopolizing a certain industry. If one company controls all or nearly all of a certain market, even if there is no illegal activity, it can be forced to dissolve into separate companies by federal regulators, as happened with AT&T in the early 1980s. While it is true that some of these antitrust laws are designed to protect consumers, there is another reason that they remain incredibly popular despite economic studies that show that industrial monopolies are always short-lived: monopolies just seem unfair.

To be sure, I do not presume to boil all of these complex social-political-legal issues down to a simple biologically ingrained instinct toward handicapping and intolerance to inequity. However, I do maintain that the fairness instinct is at least partially responsible for how we construct our social structures. I do not see it as controversial that our values are heavily influenced by our biology. Like other mammals, we are programmed to recognize inequality and be upset by it. We are programmed to seek fair play through handicapping. It seems rather obvious to me that these features of our biology would become features of our societies.

RECIPROCITY

Another manifestation of fairness and justice, as we understand them, is the concept of reciprocity. You do this for me; I will do that for you. I will give you this if you pay me with that, and so on. This forms the basis of our whole economic system—the exchange of goods and services for other goods or services of equal value, with currency and credit as handy measuring sticks for the value of those goods and services. Failure to honor this system of reciprocity is considered a crime that we call fraud or theft. Did the concept of reciprocity originate in humans, or does it predate us?

Reciprocity has been observed in a wide variety of species, including vampire bats, one of the most feared and reviled creatures in nature. These little guys go on nightly hunts to search for a nice, unsuspecting pig, cow, or horse that is fast asleep. They generally obtain their nourishment from one nightly meal, eating as much as they can without compromising their ability to fly. If they are unsuccessful on any given night, they can usually survive until the next night. However, like most flying animals, bats do not have voluminous fat deposits, and their metabolism burns fast. Two nights without eating is often fatal.

Researchers studying the feeding behaviors of vampire bats in Costa Rica observed that, at the end of the night, a bat that was not successful in finding food would often confront one that had been successful and "beg" for the other to share.[16] Surprisingly, the successful bat would often agree and feed his unlucky friend by regurgitating some of his last meal into the other's mouth. It is a disgusting scene, to be sure: one of the ugliest creatures in the world vomiting blood into the mouth of another. At the same time, it is rather beautiful: a gentle soul selflessly giving some of his food in order to save his friend from starving.

Remember that bats, like most animals in the wild, are in a constant struggle for survival, and the threat of starvation is very real. To give up as much as 20 to 30 percent of your meal means that you will be at greater risk of starving if you do not find food the next night. This is the essence of altruism: helping another at significant cost to yourself.

It turns out, however, that the altruism is not as pure and selfless as it appears to be at first blush. The bats that share food expect to be shared with the next time they come up empty after a night of hunting. You scratch my back, and I scratch yours. As it turns out, these bats remember which of their friends has previously shared with them and are much more likely to return the favor when the tables are turned. Similarly, if a well-fed bat snubs another that is hungry and begging, the rejected bat will return the snubbing if the "frenemy" comes begging in the future. This is not pure altruism; it is reciprocal altruism.

Cooperation and reciprocity have been documented in chimpanzees for decades. Scientists in the 1930s trained chimpanzees to pull a truck of food to their cages using ropes.[17] When the truck was too heavy for one chimpanzee to pull, two chimpanzees would work cooperatively to pull the truck. There are videos of this on YouTube that show the impressive task-oriented coordination that chimpanzees are capable of. But that is not the surprising part. When one chimpanzee was already well fed, he could still be coaxed into helping pull the truck by a hungry cagemate.[18] Although he certainly did not show as much determination in the task and was easily distracted, it was rather generous of him to help at all. As soon as the truck was pulled close enough, the hungry chimp took all of the food for himself without so much as a thank-you. Why did the well-fed chimp help? Probably because he knew that the tables could easily be turned in the future. It is generally a good idea to help when you can—you never know when you may need a favor yourself. Most humans understand this concept all too well.

There is no shortage of examples of reciprocal altruism in nature. Some species of birds have a specific warning call that they use to tell others that a predator is nearby.[19] By calling loudly, these whistleblowers put themselves in great danger by attracting the attention of the predator, but they do so nevertheless, even when they are looking out for others who are not even in their own family. They perform the act because they expect to benefit when someone else is on the lookout.

In chapter 4 we will discuss the existence of prostitution in animals, which could be seen as a form of reciprocity—trading sex for food or other goods. There are other kinds of "service swaps" as well. For example, Vervet monkeys will groom each other for parasitic bugs, and "returning the favor" is expected and enforced.[20] The same is true for many species of birds that will help preen each other, even if they are not related.[21]

In fact, the reciprocity of "grooming" is so common in nature that there are many examples of interspecies reciprocity that develop into full-fledged symbioses. There are species of wrasse (a small marine fish) that have evolved such that they feed purely by eating fungus and algae off of larger fish, who offer them protection from predators in return.[22] If you cruise up the Nile River, you will see crocodiles happily opening their jaws wide so that Egyptian plover birds can enter their mouths and eat bits of food stuck between their teeth. The crocodiles could easily swallow the birds in one gulp and get a nice meal out of it, but if they did, they would not get the free dental work. Similarly, you almost never see a rhinoceros in the wild that does not have an oxpecker on its back, scavenging for ticks. Scientists have documented literally hundreds of these "cleaning symbioses." What begins as a tense reciprocal exchange leads to trust and, eventually, a symbiosis. Interspecies cooperation will be discussed in more detail in the next chapter. I mention it now only to highlight how abundant the principle of reciprocity is in nature.

The reason that reciprocity is key to understanding cooperation is that the forces of natural selection function almost exclusively at the level of the individual. A genetic basis for true selfless altruism would be hard to explain because individuals prone to helping others at a cost to themselves would suffer a disadvantage and thus see their contribution to the gene pool diminish. Selfishness beats selflessness every time. However, the concept of reciprocity creates the opportunity for cooperation and prosocial behaviors to emerge. Reciprocity means that an individual will help another for self-benefit. If helping you costs me very little now but could save my hide later, I stand to benefit from helping you. In this way, reciprocity could be favored and fostered by natural selection, and a species could become increasingly social. Importantly, it does not remove the purely selfish instincts, as we will see in later chapters.

* * *

From dogs to chimps to coyotes to Homo erectus, the survival pressures of social living have led to the evolution of cooperation. Living in cooperation, however, requires that each member play her part and consume her share. Thus, the increasing socialization of mammals has brought with it the notions of fairness and equity. Over tens of millions of years, these notions have been burned into our genetic psyche. It is part of who we are, and it is not something we had to learn for ourselves.

Hammurabi's code is heralded as the oldest known system of laws, a pioneering document in our tradition of rules. Do not forget that there is little originality in that document. It is merely a work of translation. Hammurabi merely took the time to inscribe in the Akkadian language what had already been written eons before in the original language of justice: the As, Cs, Gs, and Ts of our DNA.

NOTE ON SCIENTIFIC SKEPTICISM

Skepticism is an essential component of science. It is the backbone of scientific thinking and crucial for the way that science produces knowledge. In common usage, the word "skepticism" implies doubt about something, founded or not. Scientific skepticism is a little different. Skepticism is simply the practice of requiring evidence for claims, and not just any evidence—verifiable, repeatable, and objective evidence. The bigger or more surprising the claim, the more evidence is required. If a claim is a direct contradiction of other claims that have been previously supported, the requirement for evidence is even greater.

A new idea in science often takes many years to gain support while it is tested experimentally again and again in many settings. Science is not dogmatic, and even the most widely accepted concept could be substantially weakened with a single convincing experiment. To paraphrase Richard Dawkins, a single fossil of a rabbit that dates to the Cambrian period would be enough to cast all of our ideas about the evolution of vertebrates into doubt.

When it comes to animal emotions, however, a different kind of skepticism abounds in the scientific community—that of unfounded and irrational doubt. Many scientists—and nonscientists, for that matter—simply refuse to believe that animals have complex emotions such as fairness (discussed in this chapter) and empathy (discussed in the next chapter). The studies and anecdotes that I detail in this book are all publicly available, have been presented at scientific conferences, and are published in the scientific literature. This means that they have overcome the burden of peer review, which exists to ensure that published studies conform to high standards of rigor, sound experimental design, and appropriate controls. Anecdotes and case studies must be presented transparently and labeled as such. Of course, opinions abound in the scientific press, but they must be supported by data.

While we obviously cannot access the inner experience of all the fair-minded animals in this chapter to know for sure what they are feeling, they certainly display the outward signs of fairness. People who refuse to believe or do not yet accept that animals experience complex emotions are increasingly forced to employ convoluted linguistic gymnastics in order to express their skepticism. Many scientists will say things like, "Dogs behave as if they experience empathy for an owner in pain." Then, they will launch into myriad alternative explanations for each of the behaviors they do not believe are driven by complex emotions.

It is true that for each and every example in this chapter and the next one, you could come up with an alternative explanation or two. You can almost always come up with alternative explanations when you are considering scientific observations on an individual basis. However, to support an opinion that animals do not experience empathy, one would have to come up with explanations for each of the behaviors and contexts discussed in this chapter. On the other hand, I can provide one simple explanation that unites all of the examples: animals grasp the concept of unequal treatment. So it seems science favors the simplest explanation that is consistent with the facts.

Going even further, the animal-emotion skeptics resort to pointing out that we cannot know about the feelings that animals experience, so it is pointless and wrong to make claims about them. With that logic, we should not make any claims about what other humans are feeling, either. After all, how do you know that the crying man is not just acting as if he is sad?

I will concede that when the first studies and reports of animal behaviors that indicated complex emotions started to collect in the scientific literature, the community was correct in reserving judgment. We were correct in viewing these with skepticism. I mean scientific skepticism, not simply doubt. We were correct to say, "That's interesting, but I think we need to see more data before we can conclude that these animals are really experiencing what it sometimes looks like they are experiencing."

However, now we have more data. A lot more. As you will see in the next chapter, there is a great deal of evidence that supports the hypothesis that animals feel empathy. Do we know that they do? I suppose not. But do I really know that my dear spouse is not just a mindless automaton, a holographic projection, or a hallucination? I prefer to accept that animals really do experience complex emotions, just as I accept that my spouse really does exist and fills my life with funny and interesting conversation. To put the shoe on the other foot, I would ask such a skeptic, "We have all this behavioral evidence that animals experience joy when they play. What evidence do you have that they do not?"

Let me clarify. I am not saying that the animal experience is the same as that of humans. I am just saying that animals have an inner emotional experience. While it is technically possible that animals only behave as if they have complex emotions, it is much simpler to accept that they actually have them. All other explanations seem much more convoluted and, more important, are not supported by any evidence.

Speaking of evidence, why is the burden of proof placed on those claiming that animals do have emotions? Why is the default position that humans have emotions and animals do not? Should it not be the other way around? It seems to me that, given all of the behavioral similarities documented in this book and many others, the default position should be that all animals experience mental states that we know as emotions. In my view, the burden of proof should be placed on those who would oppose that claim.

I can only conclude that people who believe that animals have no emotional experience do so because that is what they prefer to believe for any number of reasons. I do not say this self-righteously. I am not claiming the moral high ground regarding animals and their emotions.

I am, however, claiming the scientific high ground.

FURTHER READING

Bekoff, Marc, and Jessica Pierce. *Wild Justice: The Moral Lives of Animals.* Chicago: University of Chicago Press, 2009.

De Waal, Frans B. *Good Natured.* Cambridge, Mass.: Harvard University Press, 1996.

3

MORAL ANIMALS

I T IS FASCINATING to contemplate that as increasingly social mammals evolved and notions of fairness and equality emerged, a system also evolved to punish the offenders and resolve the conflict. That was the focus of the last chapter: justice. However, justice has come to imply much more than just rules and punishments. In a progressive sense, justice is also about standing up for victims of injustice. It is about righting wrongs, defending the powerless, and refusing to ignore the suffering of others.

What we are talking about here is empathy. Empathy is the capacity to feel for others, usually when they are in pain or have been wronged. It goes beyond understanding what others are going through to actually feeling something for them, as if their emotional condition transfers over to us a little bit. Most people consider this to be a very high-level human emotion. For a long time, we thought that the capacity for empathy and compassion was one of the key endowments that separated us from the animals. We were wrong. There is a long and growing list of animal species in which at least a rudimentary sense of empathy appears to be present.

AWARENESS OF THE SUFFERING OF OTHERS

One of the most striking examples of animals displaying empathy is also one of the oldest. In an experiment that is now more than fifty years old, laboratory rats were trained to press a lever in order to get food. They learned the simple behavior quickly. Then, researchers changed the setup such that when the trained rats pressed the lever, another rat was given a painful thirty-second electric shock in full view of the first rat. The traumatic effect on the observer rats was instant and lasting. After quickly understanding the new framework, they would avoid pressing the lever as much as their hunger would allow. The rate of lever pushes per day went down dramatically. Even after the electric shocks were removed, the rats remained hesitant and pushed the lever far less often than they initially had, an effect that remained for at least ten days.[1]

This study with rats inspired a similar study with rhesus monkeys. The monkeys were trained to pull on a chain that released food but also gave a shock to another monkey in their group. Because monkey behavior is much easier to visually interpret than that of rats, it was even clearer how upset the monkeys became when they saw another in pain. The monkeys became very resistant to pulling the chain. In a stunning example, one monkey went twelve days without pulling the chain a single time. He pushed himself to the point of starvation rather than hurt his buddy.[2]

As animal behaviorist Marc Bekoff frequently puts it, these experiments about empathy prove their point so well that it calls into question whether they were ethical to perform in the first place. It is pretty hard not to conclude that the monkeys and rats on both sides of these empathy experiments were in a great deal of discomfort. They were forced to choose between their own need to eat and the well-being of a comrade. In the end, they chose to eat—to live—but they did so with great trepidation. This sounds a whole lot like what humans do in the developed world. You would have to be pretty heartless not to feel guilty sometimes, living our lives of luxury and convenience in a world full of suffering. Ultimately, we still choose to live our lives, and we accept the occasional pangs of guilt that come with it.

In a gentler but even more illuminating experiment with rats, Professor Peggy Mason and colleagues from the University of Chicago found that rats will work to free other rats that are locked in a cage.[3] They do this without any reward for themselves and will help both friends and strangers. Interestingly, they are more likely to help rats of the same genetic strain, which is like a breed or race, but cross-strain empathy was also common.

Mason's group even tempted the rats with chocolate and gave them an either-or choice: free the caged rat or get the chocolate. More than half of the time, the rats elected to pass up the chocolate in order to free a stranger. Perhaps most revealing of all, the rats that chose to let their buddies wallow in captivity while they feasted on chocolate often returned to the caged rat and shared the treat with them. On average, 30 percent of the chocolate that was taken by the free rat was actually eaten by the caged rat. While food sharing is not uncommon among rodents, this almost seems like an apology.

In 2015, scientists from Kwansei Gakuin University in Japan published a follow-up to this work.[4] In this scenario, rats were also kept in enclosures and were also subjected to simulated drowning by dousing with water. (Rats have a strong aversion to water and panic when wet.) The scientists found that free rats would again pass up chocolate in order to help liberate their colleagues, and would do so much more quickly when they were in distress than when they were not. This is important because it shows that they were responding to the expressed needs of the other rat, not just the situation itself. Furthermore, rats that had been subjected to the simulated drowning themselves were even quicker to act than naive rats. These rats had "been there," which appeared to enhance their empathy.

The scientists went further and again employed the chocolate temptation. When rats were faced with the decision to stuff their faces with chocolate or help their apparently drowning buddies, they passed up the chocolate and instead helped their fellow rats more than half of the time. It seems that even rats, considered by many to be among the most vile and unwelcome creatures, experience anxiety, tension, and vicarious suffering in the presence of a fellow rat in pain. To me, this behavior sounds like it is driven by empathy.

Another major discovery in our understanding of empathy in animals was made quite by accident. Dr. Carolyn Zahn-Waxler is a scientist interested in discovering at what point human children become aware of, and show concern for, the emotional states of others. In other words, she studies when humans become able to feel empathy. In one study, her research method was to observe families with young children in their own homes (to eliminate emotional complications caused by having children in an unfamiliar doctor's office or psychological testing lab). Then, she asked the parents to, among other things, feign sadness by crying and just generally looking and acting despondent. Meanwhile, she observed the children to see if and how they were affected. By doing this, she was able to study the children's reactions and make important contributions to our knowledge of when and how empathy develops in human children.

Along the way, she also repeatedly observed something unexpected. The young children were not the only ones affected by the visible sadness of the parents. The dogs were, too. Professor Zahn-Waxler observed that, if the family had a pet dog, she frequently responded to the sadness of the parents and displayed a variety of behaviors, including those that could fairly be described as attempts to console.

Like any good scientist, Professor Zahn-Waxler did not just sit on these curious findings. She documented them, explored them further, and reported them to the community.[5] Of course, none of us pet lovers are surprised by what she found. We have all had the experience of our dogs or cats knowing when we are down and licking us or just being next to us. And it is not just sadness. Our pets are happy when we are happy and angry when we are angry. Try this at home. In a calm and stoic voice, ask your dog, "Who's a good boy?" Then, try asking the same question with a lot of energy, a high voice, and a face as bright as sunshine. You will notice a big difference. They are keenly aware of our moods and emotions, and, even further, they care about them.

To be fair, noticing that dogs and, to a lesser extent, cats show empathy toward humans is not quite the same thing as observing empathy among wild animals. These pets have evolved for thousands of years as companion animals, not just domesticated livestock. As a consequence, they experienced very strong selection for behaviors and capabilities that complement us as our steady and loyal escorts. Thus, one could argue that empathy in our pets could be an artifact of their domestication. In other words, it is clear how the ability to empathize would have benefited dogs and cats. What we are really looking for is empathy among wild animals.

Wildlife conservationist CeAnn Lambert witnessed empathy in a wild animal in her own house.[6] She found in her garage one morning that two baby field mice were trapped in a utility sink, unable to claw up its slippery walls. They were visibly exhausted and almost certainly suffering from intense thirst and hunger. They could possibly even have been on the brink of starvation, as mice need to eat constantly to maintain their extremely fast metabolisms.

Seeing the mice in this pitiable state, Ms. Lambert decided to help them escape. She gave them some water in a small, upturned plastic lid. However, only one mouse, the larger of the two, availed himself of the water and began to regain some strength and vitality in his movements. The smaller mouse still barely moved, appearing to be at death's door. What happened next was quite surprising. The larger mouse found a small morsel of food in the sink. Instead of quickly gobbling it himself, he placed it in front of the smaller mouse. The smaller mouse perked up a little bit and started to move toward the nourishing crumb. Then, the bigger mouse moved the crumb a little bit. The smaller mouse kept crawling in pursuit. The large mouse moved the food again, keeping it just out of reach of the smaller mouse.

The late Ms. Lambert was at first shocked by this apparent cruelty but eventually realized that the large mouse was leading the smaller one to the water dish! Once the small mouse reached the water, he took a drink, ate the morsel of food, and slowly began to regain his strength. Ms. Lambert placed a plank of wood in the sink, and the two mice scurried up the ramp to safety.

The take-home lesson here seems to be that mammals from rats to dogs are aware of, and moved to act by, the suffering of their buddies. While we cannot know what they are feeling, their behavior indicates empathy.

EMPATHY IN PRIMATES

Researchers in California led by Hal Markowitz trained Diana monkeys, a species of Old World monkey, to insert a token into a slot in order to obtain food. There was an elderly female in the group that did not quite get the hang of it. She fumbled with the token and was unable to put it in the slot, either because of poor dexterity/coordination or diminished capacity to understand the task. Either way, she clumsily jostled the token while the others ate. A male in the group noticed this, watched her for a few minutes, and then slowly walked over to her, took the token from her, and put it in the slot. He then walked away and allowed her to eat the food that arrived while he watched from a short distance away. This was not just a freak occurrence. The male helped his elderly mate eat on at least three occasions that the handlers were able to witness.[7]

What else could this have been but compassion? It certainly counts as prosocial or even altruistic behavior. The younger male went out of his way to help and received nothing in return. He saw a companion struggling, and so he helped.

Another heartwarming story of primate empathy comes from our closest relatives, the chimpanzees. At the Center for Great Apes in Florida, there is a chimpanzee troop with a member named Knuckles. Knuckles has the chimpanzee equivalent of cerebral palsy with a host of developmental disabilities. He has severe weakness on his left side and cannot swing or climb; he has a lazy eye and does not appear to see very well. He could not even feed himself for his first few years. He was transferred to his current home when he was two years old.

Ever since being introduced to Knuckles, the other chimps have treated him differently. They are always gentle with him. They do not punish him for spilling things or pushing into them like they would any of the other members. They are tolerant of any disturbance that he causes. Normally, a chimp his age would be subject to the usual chimpanzee dominance displays of pushing, grabbing, screaming, biting, and just general bullying. Nevertheless, the lead handler states that Knuckles has "never received a scratch."[8]

Knuckles is not ostracized for being different, either. Other chimpanzees will cuddle with him, hug him, groom him, stroke him, and lead him around by the hand. Even the alpha male of the troop has been seen grooming and caressing him. (There are videos on YouTube of the other chimps playing with Knuckles and caring for him. Go watch them for yourself. Have a tissue handy.)

Why do the chimpanzees do this? If animals are driven purely by competition for survival, how does this behavior fit? It does not. Even if we view this through the lens of social-cooperative behaviors aimed at the good of the whole group, this tolerance and compassion for a disabled member still does not make a whole lot of sense. How would the group be better off by helping a member that can never provide service to the group? Knuckles cannot repay the kindness. He can never take his place in any division of labor, and yet, the other chimps never get angry with him for this. It is difficult to see this compassionate behavior as driven by anything other than feelings of empathy.

Would the chimps show this same compassion to Knuckles in the wild, where the survival conditions are much harsher? I suspect not. The demands on their time and energy would probably be so great that they would simply have to move on and leave Knuckles behind. I do not think there is any reason to believe that they would be deliberately mean to him, but I doubt that they would expend the energy and resources to constantly support him, either. It is a rough life out there in the wild, and Knuckles would probably be too great a burden for the group to bear. (I hope you kept that tissue handy.)

However, lest we be tempted to climb up on our anthropocentric high horses, shall we ask ourselves, "Are humans any better?"

Sure, we do OK now, but until the industrial revolution, when wealth and resources began to reach the huddled masses, human existence was pretty harsh for the vast majority of humans, much as it is for wild animals. How well do you think we cared for the physically handicapped and developmentally disabled during most of our recorded history, not to mention the prehistoric and pre-agricultural period? Imagine how a rugged band of hunter-gatherers suffering through the last ice age in

Europe might have dealt with a group member with even a mild physical impairment. Do you think they would be as nice to him as the chimps are to Knuckles? It makes one shudder to think. That is, if one has empathy.

My point is that humans are really no different than the chimps in Knuckles's troop. We have the same instincts and tendencies toward compassion for our fellow humans, especially those we accept as "one of us." However, those instincts are tempered by what is feasible and practical given the needs of the group.

Our recent technological, industrial, and economic advancements have allowed us to more generously engage our natural compassion for the less fortunate. We care for each other much better now than we did when life was rough all around. It is the same with the chimpanzees. Out in the wild, the chimps' compassionate instincts are balanced with the daily struggle to survive. In captivity, however, their temporal needs are met, and they are left with much more idle time. Under these conditions, their natural tendency to be gentle and generous can manifest when faced with the strange and wonderful creature called Knuckles.

Empathy has been studied extensively in chimpanzees, gorillas, capuchin monkeys, rhesus macaques, baboons, Diana monkeys, and a host of other primates. We can now confidently conclude that empathy is a universal feature of primates. Far from being unique to humans, empathy is inherited from our primate ancestors. In all primates, including humans, compassionate behaviors are balanced with selfish behaviors, and the tension between the two is dictated by the overall availability of resources. Chapters ahead will explore this tension in more detail. Further, there is great variability among individuals in a population. This is seen in animals, just as in humans. Some are more predisposed toward empathy and compassion; others are selfish jerks.

COOPERATION BETWEEN SPECIES?

In 2003 in South Africa, a group of antelope was rounded up and captured by a private game-hunting company. The antelope were not to be killed but relocated for a special breeding program. Late one night, the workers were alarmed by the approach of a herd of eleven elephants. The herd circled the enclosure slowly and in a coordinated manner. After the brief inspection, the matriarch explored the gate of the enclosure pen and found the latches. She quickly undid all of the latches and swung open the gate. She signaled to the others in her herd, and they all retreated from the opened gate, so as to leave a free path and not frighten the antelope. The elephants stood back and watched motionlessly as the antelope cautiously approached the open gate and then sprinted through to freedom. The elephants trudged off, slowly disappearing into the night.[9]

Helping members of your own herd is one thing, but creatures organizing a rescue effort to free animals of a totally different species? That is absurd! Nature is a savage and unforgiving place where only the strong survive! As has been shown in a variety of contexts, sometimes cooperation is even more effective than competition, and many species, especially mammals, have evolved complex cooperation behaviors. Occasionally, those cooperative behaviors pop up even among different species.

Different kinds of animals working together could just be plain ol' cooperation for mutual benefit. The previous chapter showcased some examples, but there are more. For instance, coyotes and badgers have been known to occasionally hunt together as a team.[10] This is a fruitful collaboration because together they can cover the two modes of predator escape employed by burrowing rodents, their common prey. Badgers are diggers, so they can chase prey that burrows; coyotes are pouncers, so they can chase prey that scurries. Because of this complementarity, coyotes are even more successful when hunting with badgers than they are when hunting with other coyotes.

The watchman fish and the pistol shrimp are almost totally dependent on each other. The shrimp deliberately builds a nest big enough for his much bigger fish companion, who then joins the shrimp and sets up camp, and the two work together as a hunting pair. Sometimes, the fish even brings a mate along! The partnership works well: the shrimp has a weapon to immobilize prey and the fish has good eyesight.[11]

Why would animals of different species help each other? Well, why not? If they are more successful together than they are apart, both individuals benefit. It is as simple as that. But how would these partnerships have started in the first place? It is important to remember that, although some species are in direct competition with one another and others are locked in the struggle of a predator-prey relationship, many more species are simply indifferent to one another. They are in no direct competition whatsoever, so there is nothing to be gained by hostility.

Animals in close proximity often grow accustomed to each other if no threat is present. Over time, trust can build and cooperation is encouraged. I am not talking about the kind of tameness that is seen with wild animals kept in a preserve—when they seem to "know" that they cannot be hunted and are much less skittish around humans. That is just a temporary general comfort level and is not genetic or adaptive. On the other hand, if two noncompeting, nonpredator-prey species live together for a long enough period of time, it is conceivable—likely, even—that they will become genetically adapted to recognize each other as nonthreatening. From there, it is not such a big leap that they may learn to work together. If the arrangement is beneficial to both, it will be reinforced. In other words, if reciprocal altruism can benefit members of the same species, why does it not also benefit members of different species?

Cooperation is one thing, but empathy?

There are many anecdotes of cross-species empathy, most of them involving animal empathy toward humans since that is what really gets our attention. In 2009, a beluga whale saved a disoriented diver from drowning at a Chinese zoo in full view of lots of people and cameras.[12] There are two cases of zoo gorillas protecting human children who have fallen into their enclosures: Binti Jua, a male at the Brookfield Zoo in Chicago, and Jambo, a female at the Jersey Zoo in England. Both of these gorillas rescued the injured and unconscious human children that fell, gently stroked them, and showed other gestures of consolation and comforting.[13] They even aggressively stood down other gorillas that tried to approach. Why would these gorillas and this whale help humans in distress? The only reason that makes any sense is that they felt sympathy and compassion for them.

There was an elephant in India that was a trained construction helper. At one job, her task was to lift large logs with her trunk and put them into holes that had been dug. She performed this task very well with little direction until at one point, she suddenly stopped. She held the log suspended above the hole but would not drop it into place. The human workers came over to investigate the delay and found a sleeping dog in the hole. They woke it up and chased it away. Only then would the thoughtful elephant continue with her task.[14]

True empathy between different species that is not simply cooperation for mutual benefit does seem paradoxical, even to biologists who accept that animals are capable of empathy. Altruism that is directed toward a different species cannot possibly be self-beneficial and would thus be self-defeating, no?

I do not think there is a paradox here. Evolution can sometimes create sloppy or imprecise products, especially when it comes to something as complicated as behavior. Genes that allow an elephant to feel empathy and share resources with her fellow elephants may also predispose that elephant to feeling generous toward other kinds of animals that she does not feel threatened by. If animals have developed a sensory processing system that is designed to recognize suffering in their own herd, it is reasonable to think that it will sometimes be tripped by suffering in creatures beyond their herd.

Think of humans, for example. Nice people are nice people—usually to everyone and everything. Vegetarianism is more common among the tree-hugging gentle souls among us. Their compassion is just so abundant that it cannot be contained within the confines of their own species but extends to the birds in the sky and the fish in the sea. How often do you meet a genuinely nice person who turns around and beats his dog? When the genes for kindness are expressed in the animal brain, they do not necessarily include the neat instruction of "only to your own species." If otherwise gentle people are threatened or harmed, it is a different story, but their tendency toward empathy clearly extends beyond just their fellow human beings. Is the same true for other animals?

EMOTIONAL CONTAGION

Human empathy is undoubtedly complex, and scientists have been studying it for many years in all sorts of ways. The human behavior that is most closely related to empathy is the yawn contagion that occurs when one person yawns and another then "catches" the yawn. Yes, you read that correctly. Empathy is similar to yawn contagion. This is because there are neural pathways in common with these two things. Think about it: empathy is the ability to identify with other people's situations, particularly their pain and suffering. When you see or hear about another's suffering, your brain constructs those mental images so vividly that you actually cringe and feel some sadness yourself.

For reasons unknown, when you see someone yawn, the neural functioning of this "emotion contagion" is also initiated and your own yawn sequence is activated.[15] Your brain identifies with the yawner so precisely that it conjures up an actual yawn.

Empathy is kind of like that. It is as if emotional states are somehow contagious. Have you ever cried at a sad movie? Or recoiled when a boxer got clocked in the face? Or scrunched your face when someone told you about the nasty paper cut that they got? Your outward actions make it seem as though you are actually experiencing the pain yourself, even though you are not. Empathy is emotion contagion; by watching someone in pain or otherwise learning of her situation, you temporarily "catch" the pain yourself, at least outwardly.

The connection between empathy and yawn contagion was first definitively demonstrated in 2003 when researchers noticed that the higher people scored on measures of empathy, the more susceptible they were to yawn contagion.[16] This landmark study does not stand alone. Others have validated this phenomenon and probed even further. It turns out that people with schizoid disorders or antisocial personality disorder and those on the autism spectrum all display significantly reduced yawn contagion.[17] While those three disorders are all very different in origin and symptomology, they do share one feature: reduced awareness of (or concern for) the emotional states of others. There is no lack of compassion per se, particularly with autistic individuals. Rather, they do not "catch" others' emotions just by observing them, the same way that they do not "catch" yawns.

It turns out that people tend to "catch" yawns more readily from family and close friends than from strangers or casual acquaintances.[18] If we did not know about the connection between empathy and yawning, that would really be a mystery. Everyone knows that we empathize much more strongly with people that we know and love than we do with strangers—and it turns out that we catch yawns from them more easily as well. Furthermore, yawn contagion has been shown to involve similar areas of the brain as perspective empathy (considering things from another's point of view).

Finally, yawn contagion is present in other animals as well and can even jump across species. Dogs catch yawns from each other; you can catch a yawn from your dog, and vice versa.[19] The same goes for chimps and other apes. This yawn contagion is most pronounced in species that—you guessed it—display a pronounced sense of empathy.[20] I hope that I have convinced you of the connection between empathy and yawning.

Once again, emotional contagion is at the core of empathy. When you watch someone accidentally smash his finger with a hammer, you typically gasp for air, your shoulders lift, and you may even shake your hand out quickly—almost exactly like you would have if it were your hand that got hurt. But why? If nothing painful actually happened to us, why do we act as if it did?

In the mid-1990s, scientists in Italy discovered something in the brain that they thought held the answer to empathy: mirror neurons.[21] Although scientists are still studying the functions of these neurons, they appear to be an important connection between what we observe, what we feel, and what we do. These neurons are activated when you experience something and when you witness someone else experience that same thing. Then, these neurons send their information to various parts of the brain, including those that process complex emotional information.[22]

The mirror system is somehow involved in the relay of information from our senses to the so-called association areas. Association areas are where we compare sensory input to stored memories so that we can quickly make sense of that input. Once we have matched up what we see or hear with what we already know, we are able to almost experience something through someone else without having to witness it firsthand—that is the power of perspective-taking.

More germane to our discussion of empathy, mirror neurons appear to function somewhere in the bridge between the emotions of others and our own. Imagine you are at work, and suddenly a colleague receives news that her mother has passed away. Your mother is just fine. You never met the mother of your colleague, so you cannot miss her. And yet, you will certainly be moved to sadness. You may even tear up and cry with your colleague. Empathy does not pertain only to pain and sadness; it works for happy things, too. We cheer along when the protagonist of a movie succeeds; we feel genuinely happy when a janitor wins the lottery. If the mirror system is the connection between seeing others in a certain emotional state and our catching that emotional state ourselves, it could be the physical basis of empathy.[23]

What is the evidence that mirror neurons are involved in empathy? First, if mirror neurons are the key operators in human empathy, we would expect to observe some sort of malfunction in the activity of mirror neurons in individuals with autism since we know that many autistic people do not empathize as readily. Sure enough, scientists have now pinpointed some of the functional differences in people with autism, and—you guessed it— mirror neurons seem to be different.[24] Similarly, remember how contagious yawning correlates with empathy in humans and other species? Once again, the underlying connection is the action of mirror neurons.[25]

More recently, some scientists have begun to question the centrality of mirror neurons in empathy and emotional awareness. Some scientists believe that they are merely reflective of an underlying empathy system, not the engine of it. (Please pardon the pun.) While everyone agrees that the mirror neuron system is a huge discovery, many scientists are beginning to doubt that these neurons will live up to the initial hype. For our discussion here, it does not really matter if mirror neurons are, themselves, the instruments of empathy or if they are only a part of it. They could even be casual bystanders, for all intents and purposes. The point here is that we have discovered a way to measure at least part of the neurological functioning that occurs during empathy, perspective-taking, and contagious yawning. The cloud of mystery is beginning to lift. We are getting tantalizingly close to a cellular understanding of the complex emotional quale known as empathy.

If that seems farfetched, now may be a good time to remember that we are wise not to underestimate science. Just one hundred years ago, no one had any clue how sound and hearing worked, and now we have such an incredibly detailed understanding that we can surgically implant electronic devices that replace much of the machinery of hearing and send the neural-aural information directly to the brain. Biologists are now dissecting empathy with that same resolve. They have found a neural system that helps us understand what others are feeling by "mirroring" these feelings in our own brains. With that in mind, the connections between empathy, yawning, and autism are not so mysterious. If our progress in the understanding of empathy proceeds as successfully as it did with our understanding of sound, I wonder if someone will one day invent the empathy equivalent of the cochlear implant. Will we one day be able to treat people born without empathy?

It turns out that not only do other animals have mirror neurons, but we also actually understand more about animal mirror neurons than we do about human ones. The reason is that, right or wrong, the regulatory boundaries of what is considered ethically permissible are different for humans than for animals. With animals, scientists can engage in far more invasive experimentation that would be too risky or harmful to perform on humans.

In fact, mirror neurons were first discovered in rhesus macaques in the early 1990s and only later pinpointed in humans. As noted earlier in this chapter, animal empathy was first "discovered" in rhesus macaques back in the 1960s when the monkeys refused to pull a chain for food if their friend would be shocked. I find it to be a very satisfying coincidence that the same monkey species was the subject of two totally different groundbreaking experiments, decades apart, that would end up converging around the conclusion that animals exhibit empathy.

Mirror neurons have now been discovered in whales, dolphins, mice, rats, birds, all kinds of monkeys, and many other animals. Their existence seems to be widespread in the animal world, and there is an overall correlation between intelligence/cognitive ability and the prevalence and size of these mirror systems. Whether used for empathy and emotional connection or just for understanding the simple physical actions of others, there is a system in place in "higher" animals that facilitates their ability to understand the perspective of another being. This appears to be the root of empathy and compassion, and animals certainly have it.

Given that humans have more extensive cognitive abilities and more complex emotions, our mirror system would have to be quite a bit more sophisticated to keep up with all of this. Experiments have borne this out. It turns out that, in humans, our mirror neurons are not easily fooled. For example, some of the same neural circuitry is activated when you watch someone pour a glass of water as when you pour the water yourself. However, if you know for sure that the pitcher has no water in it, and the person is just faking the action of pouring water into a glass, the mirror neuron circuitry is not activated in the same way. The "mirroring" of water-pouring in your brain does not occur.[26]

This shows substantial sophistication. Most animals are easily fooled by pantomime, but humans require an elaborate ruse in order to be duped. There are two possibilities here. Either our mirror system developed complexity and intelligence hand-in-hand with our cerebral cortex, or our mirror neurons receive the information from the cerebral cortex after it has been extensively processed. Neuroscientists will surely be able to answer that question soon, but in the meantime, it does not matter for our discussion here. Human brains display much more sophistication in perspective-taking than animal brains do, and this may help explain the much more complex version of empathy in humans.

THE EVOLUTION OF EMPATHY

Now that there is convincing evidence that animals experience empathy, the questions emerge about how and why it evolved in the first place. What is the advantage to an individual to feel the pain of another? What is to be gained? The best way to understand how and why an emotion has evolved is to consider it as an internal drive to engage in some behavior. To find out why empathy evolved, we must first ask, "What does empathy drive us to do?" That will help answer our question.

Think about hunger. Hunger, in itself, is pointless; it is simply a form of discomfort. The power of hunger is that it drives us to do something: to eat. Empathy, then, is clearly a drive to help others. The feeling of pain that we experience when we see others in pain is like an internal push to help that person. It is an internal feeling of discomfort that drives us to help an injured or otherwise needy colleague. Empathy leads to compassion, which is the act of actually wanting or intending to do something to help.

In its earliest forms, perspective-taking likely emerged as a brain function for understanding the actions of others. In the simplest animals like sponges, coral, and jellyfish, the neural system is nothing more than simple input-output reflexes. There is very little processing of the inputs. However, over hundreds of millions of years, species evolved increasingly more complex neural systems. Rather than activating a simple output function, the sensory input began to be processed so that more nuanced and sophisticated outputs would result. In other words, animals with more sophisticated brains gained the ability to take in various sensory inputs, integrate them with other inputs stored previously, and build an understanding of the world around them. The neurons that are neither sensory nor motor, but instead are involved in integration and processing, are called interneurons, and the human brain is filled with them.

The number and interconnectedness of interneurons increased to staggering levels over the course of evolution. Leeches, slugs, and snails have ten thousand to twenty thousand neurons in their entire bodies, while whales and elephants have about ten billion just in their cerebral cortex, the part of the brain in which higher cognitive functioning occurs. We humans have more than twenty billion neurons in our cerebral cortex. And that is just the number of neurons. The real measure of complexity is in the number of connections between neurons. Neuroscientists estimate that humans have at least ten trillion connections, and it could be much more. By far, most of these are connections between interneurons and are responsible not for simple sensory or motor functions but for advanced processing.

As animals became better and better at understanding all of the sensory information around them, their brains developed tricks to do so more efficiently. Animal brains have the ability to "fill in the blanks" when there is only partial information. We do this by comparing incomplete sensory information with information that we have already stored as memories. It is all very automatic and occurs instantly in association areas. Imagine you are looking through a dark, crowded theater for someone that you know. Way off in the distance, you might see the back of her head, and that is all you need to identify her. Your brain fills in the rest. The whole world of optical illusions and stage magic is built on the premise that our brains will fill in the missing pieces of an incomplete picture.

How does this relate to empathy? These association areas in our brain are where sensory input is compared with what we have already seen, felt, heard, or experienced. When this happens, it conjures up those experiences in our head as though we are really experiencing them. Try this. Close your eyes and picture the face of your best friend (or father, sibling, or spouse, if you prefer). You can actually conjure up a pretty vivid image of that person in your mind. Your eyes are not taking in any sensory information about her, and yet, the visual centers of your brain are registering her likeness. This process can be observed in a brain scan. The primary visual cortex of your brain lights up when you imagine things, almost as if you are seeing them live.

In regard to empathy, your brain is wired to conjure up stored sensory imagery on command. Perspective-taking, which is the action of "putting yourself in someone else's shoes," can involve assuming someone's literal perspective, or it can involve taking their emotional perspective. Animal brains can do both. Initially, this was surely rudimentary, nothing more than interpreting the physical behavior of others through sight and sound. You would be a more successful lizard if you were able to interpret the behavior of other lizards. If you could figure out when they were seeking or receptive to mating, you'd have an advantage. If you could figure out when they were angry or threatened, that would help, too. If you understood when they wanted to play, kudos. And so on.

Empathy is the act of receiving communication, specifically emotional communication. In fact, when mirror neurons were first discovered, some scientists immediately believed they were the key to communication and language. That belief has waned, but some role for the mirror system in language and communication seems likely. This notion is edified by the fact that the most severe cases of autism involve a complete lack of language ability.[27] Regardless, there is both a conceptual and a neurological connection between emotional contagion and basic body language and signals. The genesis of empathy was the ability of animals to understand and interact with one another.

In this light, it does not take an evolutionary biologist to figure out the value of full-fledged empathy. Imagine a pack of wolves in the wild. If one member is hungry or injured and another member notices this and is moved to help him, he might just save his life. Couple this with a little reciprocity (see chapter 2), and you have a recipe for cooperation. In a social species living in tightly knit communities, members helping one another is almost always good for the survival and success of the individuals and the group. Evolution would favor this trait.

Of course, empathy, as an internal urge to do something, is always tempered and balanced against other urges. For example, if a zebra sees his friend being taken down by a lioness, he may very well feel a strong urge to help him. This urge, however, will be overwhelmed by other urges, namely the urge to save himself by fleeing. A chimpanzee may be moved to help a hungry pal, but if she is hungry herself and has only a little to eat, she probably will not help. All animals are being bombarded by senses, urges, and drives, and depending on the relative strength of each one, different outcomes are possible.

The same is true for humans. We see and hear about fellow humans in need all the time. Of course, most of us feel empathy for the less fortunate, and we really want to help, but few of us will take all that we have and give it to the poor. Our own empathy is balanced with our other drives and desires. In addition, empathy is much stronger toward those we identify as "one of us." We have all seen fundraising pitches from friends regarding some friend of theirs that has breast cancer. I bet you are much more likely to donate money when it is your own friend that is sick, rather than a friend of a friend. Do we think the stranger is unworthy of our help? Of course not. It just hits much closer to home when the subject is someone we know. This is how empathy works.

The subject of empathy often raises a very uncomfortable issue: there are people that seem to lack empathy altogether.[28] Previously referred to as sociopaths, individuals with antisocial personality disorder, or in the extreme, psychopaths, present a difficult problem for psychologists and public safety. For our discussion of animals, this begs the question: are there psychopaths in nonhuman animals? Professor Lori Marino and her students at Emory University asked that precise question and, in the process, developed a behavioral scoring system for the detection of psychopathy in chimpanzees.[29]

Yes, just as in humans, empathy is not distributed equally among chimpanzees. Some are thoughtful and some are jerks. Even more interestingly for our discussion is how similar the detection of psychopathology is in chimpanzees and humans. The Chimpanzee Psychopathy Measure developed by Marino bears striking resemblance to the way that psychologists detect psychopathy in humans. The presence of empathy in chimpanzees is made even more obvious by its absence in some members.

THE FOUNDATIONS OF MORALITY

In what little documentation we have of early human history, it seems the establishment of moral and legal codes was a ubiquitous feature of civilization, and necessarily so. If fairness is important for communities of hundreds of mammals, it will be absolutely essential for cities with thousands of human mammals. But is there a difference between enforcing equity and establishing morality? This will depend on your definition of morality and, for many, will be intensely tinged by religion. Despite the many differences among the world's religions, there is a moral common denominator: do unto others as you would have them do unto you. Christians know this as the Golden Rule, but versions of it appear in nearly all world religions and long predate Christianity.

In a purely practical sense, morality is the union of empathy and fairness. However, these two phenomena may actually be the products of the same evolutionary force—the socialization of our mammal ancestors. As Professor Frans de Waal puts it, "The two pillars of morality are reciprocity [or fairness] and empathy [or compassion]."[30] I think most of us would agree with this statement, and it certainly fits the moral framework of the world's major religions and philosophical schools of thought. (By the way, Frans de Waal is not a philosopher or a cleric. He's a member of the U.S. National Academy of Sciences who has spent his life studying the behavior of primates.)

All major religious texts contain obvious plagiarisms of writings that came before them. Despite that, all religions claim credit for giving us the code of morality. A simple look at how animals behave discredits this claim. The truth is that animals behaved according to the two pillars of morality for hundreds of millions of years before humans evolved to a state when they could arrogantly claim authorship of rather basic social-cooperative precepts as the Golden Rule. What we now consider the human moral code was written into the genes of our ancestors eons before the first prophet or shaman took credit for it. Although recent research has given us dramatic support for this hypothesis, it is not a new one. Charles Darwin himself noted it more than a century and a half ago: "Any animal [whatsoever], endowed with well-marked social instincts . . . would inevitably acquire a moral sense of conscience, as soon as its intellectual powers had become as well, or nearly as well developed, as in man." [31]

The study of the morality of animals is picking up steam in the field of biology.[32] Ethology is the field of biology that specifically studies the evolution of animal behavior. Within this field, there are many scientists that have come to the conclusion that some species of animals often behave according to a set of norms and tendencies that mirror the moral sense of most humans. They may not actively think about morality, but they behave as if they do. I find this intriguing because it is as if moral behaviors evolved before the ability to think about our behavior in moral terms.

I offer a parallel. We have an elaborate nervous system in our gut called the enteric nervous system. It not only choreographs the elaborate muscle movements that usher food through the various intestinal compartments, but it also senses the amount and composition of our stomach contents in real time. It can respond with precise combinations of a dizzying palate of secretions and hormones. The actions of all the glands and muscles in our abdomen are carefully orchestrated and responsive to the conditions from moment to moment. And yet, we have absolutely no awareness that any of this is going on. We have only the vaguest sensations of fullness and emptiness—and occasional queasiness. The rest is completely unconscious and involuntary.

My point here is that a lack of full awareness of some neural function is not evidence against the existence of that function. This could apply to the question of animal morality. Even if animals have no conscious awareness of their own moral instincts (and good luck proving that they do not), that is not evidence against their having one. Instead, the evidence that they do or do not have some moral instincts comes from studies of their behavior. Do animals behave as if they are operating on the basic pillars of morality? At this point, I think there is far too much evidence to deny that they do. This chapter and chapter 2 are filled with examples that are resonant and telling, but there are many more out there, both in the scientific literature and the popular press.

FURTHER READING

Allchin, D. "The Evolution of Morality." *Evolution: Education and Outreach* 2 (2009): 590–601.

Bekoff, Marc. "Are You Feeling What I'm Feeling?" *New Scientist* 194 (2007): 42–47.

De Waal, Frans B. *Primates and Philosophers: How Morality Evolved.* Princeton, N.J.: Princeton University Press, 2009.

4

SEXUAL POLITICS

O F ALL THE topics in this book, sex has the richest body of research
behind it. This is because evolutionary biologists, zoologists,
veterinarians, ethologists, primatologists, marine biologists,
and so on (the list could go on for pages) all spend a lot of time studying
how, when, and why animals have sex. Similarly, psychologists, psychiatrists, endocrinologists, gynecologists, counselors, sociologists—and, of
course, ministers and bishops—all spend a lot of time thinking about how,
when, and why humans have sex.

The first question you might ask is "Why do so many biologists study
sex?" There are two principal reasons for this, both of which explode into
myriad additional sub-reasons. First, sex is critically important to the life
and survival of a species. You cannot even begin to understand how a species lives without also considering how it reproduces. In fact, this is usually
where biologists start. Mating, reproduction, and all of the associated
behaviors take up a pretty significant amount of all animals' time and resources.
In fact, in many species, animals spend more time in pursuit of successful
reproduction than they do in pursuit of food, water, and other "material"
resources. Sex is such a big part of what animals do that it should not surprise us that there is so much research about it.

The second reason that sex is such a big focus of biology is that an incredible amount of sexual diversity exists in nature. It seems no two species "do it" the same way. For example, did you know that female spotted hyenas do not really have vaginas? Instead, they have a clitoris that is so large that it is technically a phallus since they also urinate through it. To impregnate a female, the male hyena sticks his penis into the penis of the female. To top it all off, these hyena females also give birth through their penises. I probably do not need to add that hyenas are a matriarchal species in which the females dominate the males.[1]

There is a bewildering number of reproductive strategies out there. You name it; they do it. And I am not just talking about mechanical arrangements of intercourse. The sex act itself is just one part of the larger reproductive strategy of a species, which involves everything from mate selection and courtship to parenting and childcare. Many books have been written about the reproductive lives of various animal species, so what I have tried to do here is point out some of the complexities of animal reproduction that seem to mirror the complexities of human sexual politics. If you think we have it bad, read on.

Part of the reason that such a great diversity of sexual strategies exists is because evolution is driven by reproductive success. The way that species evolve is dictated by which members leave successful offspring. We can summarize this (or incredibly oversimplify it, depending on your perspective) by reducing natural selection in animals to three principal factors: (1) success in surviving long enough to reproduce, (2) success in finding a mate and producing offspring, and (3) the success of those offspring in having their own offspring. The latter two of these three factors are directly impacted by reproductive strategy. This is why reproduction, perhaps more than any other feature of animals, is incredibly diverse throughout the animal kingdom. For animals with closed circulatory systems, there are just a few general designs for a heart, and all animals have slight variations thereof. A heart is a heart. The same is not true for penises or vaginas. Remember the hyena? There is a boatload of diversity out there.

WHY SEX?

The word "sex" can refer to a few different things. Of course, it can refer to the act of copulation that is very recognizable in mammals, birds, reptiles, and some amphibians. It is often called "mounting," but that is definitely not the only posture it can take. Sex can also refer to maleness or femaleness. I will discuss that use of the word and how it contrasts with gender in the next section. First, I want to talk about the basic common feature of sexual reproduction—the fusion of two gamete cells to form a zygote.

In humans and nearly all mammals, we take sexual reproduction for granted because it is all we know. There is no other form of reproduction for humans besides the sexual way. I am not saying that human reproduction requires the sexual act of copulation. That is clearly not true since the advent of artificial insemination and, more recently, in vitro fertilization. Sexual reproduction simply means the fusion of gametes—a sperm and an egg—in a process called fertilization. Whether this happens in a fallopian tube or a test tube, it is sexual reproduction because two sets of chromosomes are co-mingled, creating a distinct genetic mix.

We think of reproduction in strictly sexual terms because humans and most other mammals can only reproduce this way. However, in many other types of animals (and most plants, fungi, and protists), asexual reproduction is also possible and sometimes even the norm. Asexual reproduction can happen through some form of cloning or through parthenogenesis (the growth of an unfertilized egg into a mature adult). From insects to lizards, asexual reproduction is quite common, but very few of these species are strictly asexual. Many species can engage in either type of reproduction and will switch back and forth based on environmental conditions, time of year, population density, or some other factor.

Asexual and sexual reproduction each has its advantages. Asexual reproduction is far more efficient. For one thing, usually every member of the population can produce offspring. In a strictly sexual species like ours, only the females can reproduce. That makes the maximum growth rate twice as fast for asexual reproduction. In just a few generations, asexual reproducers can quickly overwhelm sexual reproducers. There are additional benefits in efficiency as well. For example, no time or resources are lost trying to find, attract, seduce, or capture a mate. Asexual reproducers can "do it" all by themselves. There is also no reproduction-related intraspecies competition when individuals reproduce asexually.

A common example used to demonstrate the reproductive advantage of asexual reproduction is this: if a single aphid (tiny herbivorous insect) finds herself on a rose bush in the spring, within just a couple weeks, the entire bush will be infested with aphids. Sexual reproduction, even in insects, could never give you that kind of productivity.

If sexual reproduction is so much slower and more resource-intensive, then why bother? Clearly, there must be advantages, since most vertebrate lineages have evolved to include sexual reproduction, and many, including our own, have evolved out of asexual reproduction altogether. The advantage of sexual reproduction is found in its capacity to create new genetic combinations every single generation. Sexual reproduction ensures that every individual will have novel combinations of gene alleles. Those aphids colonizing the rose bush will all be clones. On the other hand, consider yourself and your siblings. You are probably a diverse bunch, even with the limited gene pool of just your parents. That is the power of sex. By tying reproduction to genetic reshuffling, diversity is ensured over the generations. Even with only the same two parents, a couple could have trillions of children and no two would be the same genetically, save for any identical twins.

Why is genetic diversity good? At any one point in time it may not be, but over the long term, it is essential for the survival of the species. It may be true that the aphids on our rose bush are perfectly suited for their habitat and lifestyle. They may be so well adapted, so successful, that any change, however slight, would be detrimental. In that case, cloning makes really good sense, and that is why they do it. However, that will only be true for as long as the environmental conditions remain the same and resources are plentiful. By autumn, the rose bush may be dropping its leaves. Or maybe the bush gets so badly eaten that it dies and there are only grasses and sunflowers nearby—and maybe those sunflowers already have insects established there. What do the aphids do then? They must find some other way to live. They must adapt. And guess what? While aphids prefer cloning in the springtime, they inevitably switch to sexual reproduction in the autumn, which helps them prepare for new challenges, especially the harsh winter ahead.[2]

Adaptation can only occur if there is a wide diversity of individuals because individuals do not adapt; populations do. The species as a whole adapts through the differential survival or death of its members. If everyone is the same, everyone lives or dies the same way. If everyone is different, then there is hope that at least some members of the population can survive whatever the next challenge is. Sexual reproduction prepares a species for an unpredictable future. The environmental conditions of our world, both biotic and abiotic, are ever-changing. Even the most successful clones have a bleak future when things change in their world. This is why almost all animal species are capable of sexual reproduction, even if they do not always do it.

Sexual reproduction has evolved as the most successful means to ensure the long-term survival of the species, despite its many drawbacks. It is slower and more costly, as we said, and it also leads to the sometimes absurd phenomenon of sexual selection, where features evolve and are selected purely because they provide advantages in sexual success, not necessarily survival. For example, there is a species of Drosophila that produces sperm that is twenty times the length of its body; its testicles make up more than one-tenth of its body mass.[3] That may be good for reproduction, but it does not really help with survival. We all know the obscene size of the peacock's tail. Antlers are a rather ridiculous thing to have to lug around all year, and so on. Despite some of these evolutionary dangers, sexual reproduction has thrived on our planet because of its unique ability to provide a species with a rich genetic toolkit for surviving in a changing world. Simply put, sex is good for us.

SEX VERSUS GENDER

The terms "sex" and "gender" can be confusing because they mean different things to different people in different contexts. In terms of human psychology and sociology, sex is biological and gender is a sociological construction. Gender identity is one's understanding of one's own sex, gender, and sexuality. Gender expression is how we live out our sex and gender in the world. This includes clothing, hair, and makeup, as well as sexuality. It is complicated, and I think it is pretty obvious that we humans have our own impressive sexual diversity.

Still, the diversity of human sexuality pales in comparison to that found in animals. When discussing animals, the terminology is a bit different, and it is currently changing as we gain more appreciation for all the complexities involved. There is but one commonality throughout the entire animal kingdom: for all species, there are only two possible gametes—a big one, called the egg, and a small one with a tail, called the sperm. Sexual reproduction occurs when new individuals are formed through the fusion of a sperm cell and an egg cell.

For animals, biological sex refers only to which gamete is made—egg (female), sperm (male), or both (hermaphrodite). Yes, hermaphrodites are very common in the animal kingdom, particularly among invertebrates. The term "gender," on the other hand, was not used for animals until very recently. I am a fan of the definition of gender recommended by biologist Joan Roughgarden: gender refers to the particular reproductive strategy employed by an animal.[4] Of course, sex is a big part of gender, but gender is more than just one's sex, in both humans and animals. In animals, gender refers to courtship, mate selection, copulation, pair-bonding, nesting, caring for children—the whole spectrum of reproductive behaviors. We speak about animal genders when we describe their behaviors in these areas. While biological sex is usually rather simple (male, female, or hermaphrodite), gender can be complicated. In many species, there can be multiple genders within a single sex. There can be different "kinds" of females or males that are fundamentally different from each other.

This is best illustrated by example. There is a species of European wrasse in which scientists have discovered multiple male genders (but just one female gender). European wrasse are small, colorful fish that live mostly in the Mediterranean Sea. In wrasse, the different genders of males are true males—all make sperm and can sire offspring with those sperm. However, each male gender has a different strategy for "fathering" offspring. Wrasse females do absolutely no parenting of any kind. They simply squirt eggs into a nest prepared by a male and are gone forever. The males, however, spend weeks building and preparing the most elaborate and protected nest in order to entice the females to provide eggs. Then, they guard the fertilized eggs vigilantly for another week until the baby fish hatch and swim away. Right away, we can see that any preconceived notions of anthropocentric gender roles should be quickly dismissed.

However, not all male wrasse are the same. Some are the "typical" large and strong males. They do all of the work of building the nest and most of the work protecting it. They chase other males from the nest and show hostility to territory encroachment. Some males, however, are born different. They are much smaller and have nothing to do with the other males, or the females, for that matter. They do not build nests, they do not guard them, they do not protect eggs, and they do not care for young hatchlings. However, they still manage to father a fair number of offspring. They do this by being sneaky fuckers. (That is a real, if unofficial, term in evolutionary biology; I am not just being crude.) These little wrasse males have a very simple reproductive strategy. They hide in the vegetation and watch for females to deposit eggs in a nest. Then, they swim in as fast as they can and squirt their sperm before being chased out by one of the larger males. It is sneaky, but it is effective enough to father children at a respectable rate. A particularly sneaky fucker can fertilize eggs in many nests, which means his genes will mix with that of many females. It is not a bad strategy.

We are still not done with these interesting fish. It turns out that there is yet a third gender of male wrasse. These males are intermediate in size and share some markings with females. They are known as "helper" males because they offer partnership to the large males with guarding the nest both before and after the female deposits the eggs. They act as lookouts and chase away other large or small/sneaky males and possible egg predators as best they can. If the large male accepts this help and enters into the partnership, he will agree to let the helper male fertilize some of the eggs with his sperm when the eggs arrive. Then, if the deal was honored, they stand guard together for a few days. It is a model of cooperation for mutual benefit, proving that mammals do not have the monopoly on cooperation. Fish can do it, too, and not just in mindless collective schools, but as individuals.

If you are thinking that this one species of wrasse is just an aberration—a weird species that is like no other—think again. Many species of fish have two or three genders of males. In fact, one of the most abundant fish in freshwater ponds and lakes throughout North America, the bluegill sunfish, is one of the best understood of the fish with three male genders. Experienced anglers can easily tell the difference between males and females. Or so they think. Many of the fish they identify as females are actually helper males. In sunfish, the alliance-forming ritual between large and helper males is a courtship dance that includes genital contact. Quite often, when a female joins the picture to contribute the eggs, the sex is a three-way affair.

The cooperation of large and helper male sunfish provides an interesting example of same-sex parenting, but more on that later. I mention these examples here to explain how the term "gender" is used for animals. It is different than sex. Gender in animals defines the way an animal lives and behaves in ways connected to reproduction (as opposed to, say, hunting). Most fish do not copulate, at least not in the way that we think of it. Separately, the two sexes squirt their gametes into a nest. However, they do have a "sex drive" in the sense that they engage elaborate strategies that will maximize their reproductive success.

The three genders of wrasse and sunfish males explained here are not just lifestyle choices or the result of some male fish being bigger than others. Fish are born into these genders in an unambiguous way, and two of the male genders have coevolved this cooperative behavior. This is a gender, a hardwired property of the individual that determines reproductive style and capacity. Although this gender is controlled by genetics, it is manifested in behaviors. Thus, this is a good example of complex behaviors being controlled by genes and proteins. The fish do not learn these behaviors from their parents, nor is there an apprenticeship program for learning the trade. The males just know what to do based simply on their gender. The male genders are kept in balance by natural selection: if one of the partnering genders becomes too abundant, they will become less successful because they will not have partners.

Slight variations of the multiple male gender patterns are seen in many other animal species. For example, there are two genders of male red deer, one with antlers (stag) and one without (nott). One could write whole books on this phenomenon. And, in fact, many already have. (I heartily recommend Joan Roughgarden's book *Evolution's Rainbow* if you want to read more.[5]) The summary of this is that the particular reproductive strategy that an animal exhibits—its gender—can go much beyond its sex. And is not the same true of humans?

MUCH MORE THAN PROCREATION

The largest misconception that we must dispel about sex is that animals do it only for the conception of offspring. This incorrect notion is the biggest hindrance to understanding animal sex and, unfortunately, is the most widespread. Even many scientists who are not field biologists think this way. We will see that animals use sex for establishing and strengthening pair-bonding, for greetings to build familiarity and trust, for breaking tension and thus reducing aggression and violence, and for establishing dominance hierarchies in a social group. If you think that humans are the only animals that have sex for any reason other than procreation, think again.

Truth be told, humans do not even take the prize for the most sex-obsessed animal species—not by a long shot. Bonobos, which are among our closest relatives, are a species notorious for their constant sexual escapades.[6] Quite literally, sex is like a handshake, a greeting. Sex is used to resolve disputes and conflicts and again when the conflict is over. Humans are not the only ones that enjoy makeup sex. Bonobos engage in tongue kissing, oral sex, and all manner of mutually pleasurable rubbing. None of these are procreative. They even get aroused when they find food. Often, if a group of bonobos happens upon a large new source of food, their excitement will be too much to bear, and they will engage in a hurried orgy before they dive into the food.[7] The resulting sexual afterglow ensures a peaceful meal without petty squabbles over portion size.

Recent research has revealed an important qualifier for the sexual reputation of bonobos: they are much more sexually active in captivity than they are in the wild, and there may be other differences between captive and wild behaviors as well.[8] However, I do not think this dooms our comparisons to human behavior by any means. We also generally have more sex when we are not busy or stressed.

There are a variety of ways to disprove the myth that animals copulate only for procreation. For example, in most species of mammal, females have a narrow fertile period in their estrus cycle that is sometimes called a heat. Only during heat will she release a mature oocyte capable of fertilization. However, this does not mean that she will only engage in sexual activity during heat, nor does it mean that males will only seek sexual activity with her during this period. Sure, in most species, females engage in more sex during heat and males are more attracted to them when they are in heat, but they still have plenty of sex outside of heat. Why would that be?

In the past, some assumed that animals were just too daft to know the difference between when they were fertile and when they were not. The claim was that the sex drive is always there, but it is just meaningless when the females are not in heat. There are two problems with this claim. First, there is no evidence for it. It may seem like a reasonable assertion, but scientific claims are not based on reason alone—they must be supported by evidence. A claim that lacks evidence is tentative, at best. Second, there are mountains of evidence for nonprocreative roles for sexual activity in all sorts of animals, some of which we will soon explore.

By the way, it has been well documented that human females have a slightly stronger sex drive during their fertile period. Another time when human females have a stronger sex drive is when they are pregnant. It does not take a biologist to guess the evolutionary benefit of that, and it most definitely is not an effort to conceive again.

Before I launch into the many functions of sexual activity in the animal kingdom, I cannot gloss over what I think is the reason why so many people, including some scientists, believe that animals have sex only to reproduce. I think it is at least partly because they are projecting their moral values onto animals and thus prejudicing their scientific opinions based on their feelings about human norms. It is a form of upside-down moral anthropomorphism. Because animals only have sex for making babies, humans should only have sex for making babies. So goes the reasoning. The problem is that it is just not true. There is way too much evidence to hold that belief anymore. You can believe what you want about when and how humans should have sex, but to believe that animals have sex only for procreation is pure fallacy.

THE PLEASURE OF SEX

As we begin, we need to keep in mind that the predisposition toward having sex, the sex drive, is hardwired in two ways. First, there is the urge itself, which requires little or no training or prior experience. In other words, there is an inborn instinct. Then, there is the pleasure-reward system. This involves the release of neurotransmitters in the brain that we experience as "pleasure." There is a pleasure center in the brain that is sensitive to these neurotransmitters, which are released when we experience something pleasurable. Neurotransmitters hitting the pleasure center is what pleasure is. Long ago, evolution invented this technique as one way to drive animals to engage in a behavior that is beneficial for the survival of the species. Once you experience this pleasure, you will want to experience it again. It is like a drug—very much like a drug, it turns out. I am not saying that animals, especially those with much simpler brains than ours, experience pleasure the same way that we do. But they clearly have instinctual drives and urges to perform certain behaviors.

It is fair to say that the origin of the pleasurable nature of sex was probably to promote reproduction and that the "original" purpose of sex was to bring together sperm and eggs. I say it is fair, but not that it is correct, because we do not know that for sure. It does seem likely, since that aspect of sex is "older" than other functions of sex and is retained as a universal function of sex. Nevertheless, we cannot be 100 percent confident because we know very little about the behavior of ancient animal species. Behaviors do not fossilize as easily as bones do.

None of this really matters to our discussion, however. The fact that sex may have been originally for conception does not change the fact that it is now used for much more. It is like the wings of a bird. In the reptilian ancestors of birds, forelimb structures were not used for flight. Does that matter to present-day birds? Are they are using their forelimbs in an unnatural way? Of course not. Animals evolve new uses for old things. That is what we do.

In animals, the "pleasure response" evolved to promote certain behaviors (usually instinctual, but not always) by rewarding those that perform the behaviors with the sensation of pleasure. Thus, it was only natural that sex would be among the first behaviors to get the honor of being pleasurable to the animal brain. This pleasure response would have made the first animal to experience it become slightly addicted to having sex and thus leave lots of offspring that would then have the addiction, and so on. Over time, natural selection would enforce a balance between the addiction to sex and the need to do other things in order to stay alive and be successful.

However, once sex and pleasure were linked in the brain, this was a powerful tool for nature to work with by using sex for other beneficial purposes. If the brain was already wired to enjoy sex and seek it out, all that was then required was the right social setting in which other purposes of sex could then naturally emerge. Once again, it is like the front limbs of the amphibians and reptiles: once this structural scaffold was in place, the limbs could then be used for different things. In some descendants of the reptiles, the forelimbs evolved into wings. The advantage of having wings is obvious, and the developmental stages along the way from limbs to wings must have had advantages, too.

The evolution of a behavior is a little different from that of an anatomical structure, but the principle is the same. The phenomenon that sex is pleasurable was the scaffold onto which other sex-centered behaviors could emerge. If the act of sex ever became helpful for something else entirely—say, building alliances with higher-ranking members of the group—there would already be a built-in way to encourage that: pleasure. Keep this "pleasure principle" in mind as you read the pages ahead. Animals have sex because they enjoy it, and it is possible that in some cases, the pleasure is all they really "understand" about the sex that they are having—they do it because they like it. But that does not mean that it cannot also serve some other function of which they are not necessarily aware.

TRICKY USES OF SEX

Some animals use sex as a weapon! Competition for resources and reproduction is fierce, so there is great pressure to use everything in your arsenal against the competitors. For example, there is a species of garter snake in Canada that enjoys wild group sex when they emerge from hibernation in early spring. What happens is that the males tend to emerge first and then wait for the females. As each female emerges from her burrow, a group of males will surround and entangle her in a large swarm of sexual frenzy. It is a wild sex orgy, and who can even keep track of who is who?

However, some males will emerge from hibernation emitting female pheromones.[9] This is sometimes called female mimicry and is the snake equivalent of Bugs Bunny putting on a wig and lipstick. What happens when one of these garter snakes emerges from the tunnels smelling all sexy and ladylike? You guessed it: the hordes of waiting males surround him and do their thing. There are no external genitalia in snakes and no sex-differential markings in this species, so the males are easily fooled. But why would he do this? What does he get out of it?

Snakes are exothermic (cold-blooded). They do not generate their own body heat; they have to get it from the environment around them. By putting himself in the middle of this orgy, a male can steal body heat from the other males. The fact that there are males waiting for him means that he was not the first to emerge and is already behind in the race to inseminate the females. He is also cold and sluggish when he emerges. What better way to warm up and get up to speed with your competition than by tricking them into putting you in the center of a giant bear hug? Once this cross-dressing snake is sufficiently warmed up, he will take his place among the other males in the competition to inseminate the emerging females.

In addition, it may also be that the female-mimicking garter snake does this in order to distract or exhaust competitors. By pretending to be a female, he tricks the other males into using up energy in a futile effort. This way, he will have an advantage over them going forward because he is fresh from hibernation and raring to go. The point here is that some male snakes clearly trick other males into having sex with them, and they gain a competitive edge when they do so.

An even trickier use of sex has been discovered in flour beetles. Beetles, like most other insects, are known to frequently engage in same-sex mounting. There are many possible reasons why this might be so. Among them, it has been suggested that a mounting male beetle may actually be depositing sperm inside the reproductive structures of another male in the hopes that the "receiver" beetle will then subsequently transfer that sperm, along with his own, when he mounts and inseminates a female. Sounds far-fetched? Sara Lewis's research group at Tufts University found that this indeed happens. They found that males that mounted other males often ended up fathering a small but consistent portion of the offspring of females that the second male subsequently mounted.[10]

This is quite an effective strategy because, by mounting other males, a male can potentially spread his sperm to not only the females that he mounts, but the females that other males mount as well. Pretty sly.

Lest anyone get the impression that same-sex sexual actions have the monopoly on manipulative sex, consider sexual cannibalism. Mostly found in spiders and a few insect species, this practice involves the female consuming all or part of the male, prior to, during, or after sperm transfer.[11] The cannibalism thus provides a large amount of calories and nutrients to sustain the offspring of the gruesome mating. That is all fine and good, and if that is the understood life cycle of the species, so be it. Tough luck, guys. At least you would die knowing that your children will be well provided for, which is of course the adaptive explanation of why males willingly succumb.

Not so fast. In many species, females are known to occasionally "fake it." Either by brute aggression or by tricking a male into giving up the ghost voluntarily, some females will consume a male without laying eggs.[12] Instead, she continues on her search for a "better" male, one that is not so easily caught or tricked. Harsh.

This type of trickery can even work across species. New Zealand scientists have noted the decline of the native species of praying mantis as foreign praying mantises were introduced. This is called invasiveness in ecological terminology. The invasive species of praying mantis seemed to be out-competing the native species very quickly. When scientists took a closer look, they were shocked by what they found.[13] The females of the invasive species emit pheromones that strongly attract the males of the native species. Horny to the point of disorientation, the males humbly offer their lives in exchange for sex with these attractive new females. The females happily gobble them up and keep moving. The females of the new species did not bother laying eggs, which would not have been compatible with the sperm anyway. They just took the free meal and continued their invasive march.

Throughout many species, we can see sex being used for things other than simply fertilization/conception. I do not think I even need to mention that the same is true for humans. Many people use sex, especially sex appeal, to manipulate others. Have you seen a car commercial lately? Some people use sex for revenge. Others use sex to shift the power balance in an otherwise nonsexual relationship. Lots and lots of people market access to their sexuality as a commodity. Between prostitution and pornography, sex is a multibillion-dollar industry. I am not defending or promoting any of this, nor am I condemning it. All I am saying is that these sundry uses of sex are widespread and in no way to unique to human beings.

In humans, as in animals, sex is a behavior. Behaviors are just like anatomical parts: they can be utilized in a variety of ways far beyond whatever their "original purpose" may have been. Natural selection will reward creative animal species that shape their anatomy and their behaviors into new and clever purposes. At the same time, libido is a strong instinct, a behavioral drive, and this creates a potential weakness in an individual that others can exploit. Animals, including humans, will go to great lengths to satisfy their lust. It is no surprise, then, that clever animals will use the sex drive of other animals against them. Of course, humans do the same. After all, we are animals (especially when it comes to sex).

SEX AS A TRANSACTION

The discussion of prostitution, pornography, and the sex trade introduces the concept that sex can be transactional. Humans did not invent that concept. Far from it. Given the natural drive for sex, it was only a matter of time before some of our animal forebears would trade sexual access for food or other goods.

There are many species of spiders and insects in which one sex, most often the male, will present food as an offering in order to gain access for mating. Females can then be choosy about who brings the best offering.[14] This has been seen in birds as well.[15] Males sometimes have to "purchase" access to females or, more bluntly, to their eggs. However, researchers studying penguins in the Antarctic found that pair-bonded females will have extra-pair copulations (EPCs) in exchange for rocks, which are used to build nests.[16] The point is that female penguins will trade sex for a commodity. This behavior is only typically seen when rocks are scarce, leading to an interesting headline in The Style Forum: "Even Penguins Prostitute in a Recession."[17] This behavior comes quite close to the human definition of prostitution because the sex is not part of a courtship or a pair bond, it is a quick one-time thing, and it is exchanged for a tangible resource.

In macaques, sex in exchange for grooming has been noted. Sex for meat has been observed in chimpanzees. In one very interesting example, capuchin monkeys held in an animal research facility were trained to consider small silver discs as a form of currency to purchase food or other favors from their handlers. Before long, some of the males were seen giving these token to females in exchange for sex.[18] It seems that prostitution really is the oldest profession.

In humans, trading resources for sex is not always as simple as professional prostitution. How many piggish men do you know that downright expect some sexual gratification after spending lavishly on a date? I am not condoning that, but it is unfortunately very common, and I think most would agree that these men are behaving in an uncivilized—one might even say animal-like—manner.

This phenomenon has been studied on college campuses, and the results will probably depress you. In short, 27 percent of men and 14 percent of women willingly admitted that they have attempted to "purchase" sexual access to a potential partner.[19] This is rarely as blunt as a direct offer of money. It is more in the realm of trying to impress with fancy gifts, expensive dinners, elaborate favors, helping with homework or studying, and so on. Perhaps more surprisingly, most individuals in this study reported that they recognized when the niceness of others was actually an attempt to purchase sexual access. Even further, at least 25 percent of the time, the solicited party honored the mutually understood arrangement and "delivered" on his or her end of the deal. At least on American college campuses, there is a vibrant economy for the sex trade.

Even more interestingly, this study revealed that the opposite transaction also occurs. Five percent of men and nine percent of women reported having attempted to give sexual access in order to get something. This is something like, "I'll sleep with you if you do my laundry/give me your Xbox/buy me that purse."

The point of this study was to document what we already know all too well. For humans, sex is frequently a commodity. It is often included in social and economic transactions, and some people are more effective at using it to their advantage than others.

The transactional nature of sex is not always just about sex for pleasure. Procreation can be part of the equation as well. Just like with fish and spiders, we all want to reproduce with someone who has a lot of wealth (more on this in chapter 8). The mingling of wealth and procreation is indeed something we share with animals. Just look at the various customs, now thankfully beginning to fade, surrounding the marriages of aristocrats, nobles, royals, and tycoons. One of the best ways to improve the standing of one's family is to marry into a family of higher station. But that will cost you. After all, what are you offering to them that would entice them to "marry down?" Swirling within this retched snobbery is the bald truth that reproductive access is a tool, a bargaining chip with which to achieve social advancement.

Never is the transactional nature of reproductive access made clearer than in the phenomenon of arranged marriage. In most of the modern Western world, we react with horror at the very concept of arranged marriage. Yet, in our smugly enlightened state, we forget that for all but the last three hundred years it was almost universally practiced across the globe, and it still is in many parts of the developing world. The customs vary widely regarding who does the choosing, who has veto power, and all of that. I do not want to veer too far off the rails here, so suffice it to say that regular old autonomous marriage for love is definitely the historical anomaly for our species. For most of our civilized history (and likely much of our prehistory), we have viewed marriage and procreation through the cold hard lens of cost-benefit calculations.

The dowry is another widespread practice in which the parents of the bride bequeath a large sum of their family resources to their daughter and her new husband upon their marriage. The specifics vary, but the size of one's dowry has historically been a very substantial aspect of a young girl's attractiveness to potential mates. This practice was common in the ancient empires of Europe, Africa, the Americas, the Middle East, India, China, and Southeast Asia (so pretty much everywhere). The dowry has a reciprocal as well: the "bride price," which was almost as widespread. The underlying theme of the dowry and the bride price is the concept that procreative access is a resource, a commodity, and can be purchased in, or at least associated with, economic transactions.

This brings us full circle to the earlier example of fish and spiders that must "purchase" reproductive access with food or other goods. In nature, this is called a nuptial gift, and it is quite common. It is typically a male that must provide something in exchange for access to a female's eggs. When the nuptial gift is a food item, this is often called courtship feeding, and many spiders and insects engage in this behavior. A bird called the great grey shrike must provide a meal for a female to allow him to copulate with her. This meal is a recently killed prey animal, and size matters. Interestingly, these birds forms pair bonds, but paired males will continue to purchase sexual access from other females.[20]

In a rather disgusting example, male katydids actually produce a secreted gelatinous food substance and offer it to the females.[21] Packed with calories and nutrients, this nuptial gift, if accepted, is traded for sexual access. It is rather like cooking a very elaborate dinner for your potential wife in order to impress her before popping the question. Except in this case, the meal is made of one's own bodily secretions. And it gets worse. Tucked inside the meal are the sperm cells of the male. The nutrients and sustenance of the spermatophore, as this gift is called, are absorbed right into the female's genital canal. Once the outer package of food is absorbed, the sperm cells are free to carry out their mission.[22] This proposal-night dinner is more like a Trojan horse.

MASTURBATION

Many people find it surprising to discover that animals masturbate, which does not really make any sense since we have all seen a dog hump a pillow, a stuffed animal, or a leg. My guess is that our reluctance to see animals this way is rooted in the shameful way that we regard human masturbation.

The reality is that masturbation has been documented in many animal species, including almost all mammals.

Some of the species in which masturbation has actually been studied include bears, walruses, lions, cows, dogs, cats, horses, donkeys, kangaroos, and even porcupines (ouch!).[23] Orangutans have even been known to use toys such as makeshift dildos (to penetrate) and fleshy fruits (to be penetrated).[24] Also, it turns out that just about any animal that is physically able to reach will self-stimulate orally from time to time. Animals are well known to rub against objects to obtain sexual gratification. For example, moose and caribou achieve sexual gratification by rubbing their antlers on trees.[25] There is a video on YouTube of a chimpanzee using a frog to masturbate and another of a river dolphin masturbating with a dead fish.

In most species, animal masturbation leads to orgasm, although it varies widely how often it does. For example, in dogs and horses, masturbation only rarely leads to orgasm,[26] while in cows and our fellow apes, it almost always does.[27] Female animals do it just as much as males do, although often differently. In many mammals, females can masturbate through imitating the "male behavior" of mounting and then rubbing their genitals on whatever they are humping. Females will do this with males, other females, or even inanimate objects. Bottom line—animals masturbate in a wide variety of ways.

Why do they do it? Well, why not? As we have already discussed, evolution has linked up sexual activity to the pleasure centers of our brain as a means to encourage it. Sexual stimulation, especially climax, feels good, even if it is done all alone. For that reason, if no other, we should expect that animals are going to masturbate. And why not, if no harm is being done? As usual, nature provides a more interesting story than that. It turns out that there are several good reasons to believe that masturbation is not just neutral but is actually good for individuals and species.

The first guesses about the benefits of masturbation came from within the larger idea of "sperm competition" in many animals.[28] For species in which the females mate with many males, he with the most sperm has a certain advantage when it comes to acquiring paternity. This is why, so the thinking goes, there is a rough correlation in primates between the promiscuity of the females and the size of the males' testicles.[29] From this, some have concluded that "cleaning out the pipes" would be a good idea to be sure that no old sperm were lying around. The thinking here is that routine clearing out of all of the sperm-factory plumbing raises the overall quality of the sperm count. Further, it may be beneficial to regularly clean out the tubes and glands that make the seminal fluid. It makes sense that, if you want sperm to perform at their best, the accessory juices should be as fresh as possible.

This explanation of the value of masturbation is supported by a couple of studies, including in macaques, a primate relative of ours.[30] However, I am not convinced that this is a widespread function of masturbation. For one thing, does the presence of some old and slow sperm really have any negative effect on the new and fresh ones? If so, how? Also, sperm can only be made so fast. Even if it were true that masturbating every so often clears out old sperm, would there be a fine line past which masturbating too often amounts to tossing out perfectly good sperm and thus reduces the sperm count of the next ejaculate? Also, this fails to explain the masturbation of those species that do not often orgasm from it, like horses and dogs. And what about the females? This does not say anything about them.

Professor Jane Waterman published an article in 2010 detailing the masturbation habits of a species of African ground squirrel.[31] She studied twenty males and found that all twenty of them masturbated regularly. They masturbated more when the females were in estrus (i.e., heat), which could be due to the sexy mood caused by all those pheromones floating around. Curiously, males tended to masturbate after sex (as opposed to instead of) and the more dominant a male was, the more he masturbated. This seems quite different from what we would expect. We usually think of masturbation as something someone does instead of sex, as a substitute when we cannot get the real thing. Not so with these squirrels—the dominant males have the most sex with the females and also do the most masturbating.

Professor Waterman hypothesized that the masturbation was a form of hygiene, a way to clean the genitals, inside and out, to prevent infections, including and especially sexually transmitted ones. Sexually transmitted infections (STIs) are a real problem for these squirrels because they reduce fertility. (They are generally not fatal because the squirrels have very short life spans anyway.) During a mating season, these squirrels enjoy one three-hour sex romp during which a typical female will mate with ten males. A nasty STI would spread through the population very quickly and potentially render them all infertile. Thus, urogenital hygiene is a serious matter—not just for the reproductive success of an individual but also for the survival of the group.

One more thing about these squirrels. In all of the 105 masturbation sessions that were observed in this study, the males consumed the ejaculate that resulted. There are two likely reasons for this. First, the mouth is used during the masturbation/grooming because of the antibacterial properties of saliva. Second, the ejaculate is consumed in order to avoid needless water loss. These squirrels are well adapted for the dry climate, and the consumption of their ejaculate seems to be part of that adaptation.

Moving on. One reason that some female mammals masturbate is because it is necessary to maintain their reproductive health. So, some females simply must masturbate. The reason is that some species have what is called induced ovulation, where an egg is not released until the female is properly stimulated. This is a very clever arrangement, especially for species that are accustomed to living in low population densities. If ovulation happens on its own schedule, an oocyte will be wasted if a male is not around to fertilize it within a day or so. While sperm can usually live for a few days in the oviducts, oocytes are much more fragile. It makes good sense for females to release the eggs only during sex. Some induced ovulators are cats, ferrets, rabbits, camels, and llamas.

In these animals, when an oocyte is ready, it can hang around inside the follicle for a little while, but it must eventually be expelled from the ovary or it could develop into a cyst. In some species, for example ferrets and cats, a ripe egg is actually painful if not ovulated within a few days. Anyone with an intact female cat can tell you that when they are in heat, it is obviously a painful experience if they do not have sex. Veterinarians and cat breeders know how to relieve this pain. There are products on the market specifically for stimulating ovulation in cats. They resemble Q-tips. So, it is not a surprise that female cats and other induced ovulators have figured out how to relieve the pain all by themselves: through masturbation.

Another idea about the value of masturbation in male animals is that it hones the sexual prowess of the animal by giving them "practice" as to how to mount and thrust properly. This is similar to the "play as practice" theory discussed in chapter 1. The idea is that, as males become sexually competent, the instincts toward humping and mounting emerge and are developed through practice.

Do any of these hypotheses apply to humans? Well, it turns out that regular ejaculation actually does improve the quality of human male sperm.[32] This seems to be because the longer that fully mature sperm are held in storage, the worse they will perform when they are unleashed. This makes sense because, to conserve their energy and stamina, sperm are stored with activity inhibitors in the epididymis, a meshwork of tubes on the back of the testes. These act as sedatives to keep the sperm sleepy. Then, when the sperm are released and join with the other components of semen, they are hit with stimulators and a rush of energy sources such as fructose and citric acid. Think of the seminal fluid as Red Bull. When you look at it like that, it is not altogether surprising that the longer the sperm are held in a sedated state, the less likely they are to simply perk up and go to work when the sedatives are reversed. Essentially, this is the human version of the sperm-competition phenomenon, where the benefits of masturbation are found in the enhanced quality of the sperm.

As we attempt to discover various biological benefits of masturbation, it is important to remember that it is not important that we do not, nor do all those animals, have any realization about the benefits. From our perspective, we simply do what comes naturally, what feels good. This brings us back to the "sex feels good" phenomenon and the tendency of nature to find multiple purposes for an evolutionary innovation. As with all traits and behaviors, it is likely that there are many overlapping explanations for why animals masturbate. I suspect that each of the benefits of masturbation noted in this chapter contributes to the prevalence of masturbation in some species and contexts. I also suspect that there are additional underpinnings that we have not even guessed at yet.

However, it may be that animals and humans masturbate for no other reason than that it feels good. Biologists have never observed any harm coming to animals because of masturbation within reasonable moderation. Certainly with humans, psychologists and physicians have found no harm. In fact, there is a great deal written about the benefits of masturbation for the normal psychosexual development of teenagers.

The bottom line is that masturbation stimulates the pleasure centers of the brain that have been conveniently linked up to sexual activity. Because of that, we are driven to satisfy sexual urges and seek that kind of pleasure. Thus, masturbation would naturally have emerged no matter what. It could even be that there are no benefits whatsoever, and yet masturbation is still likely to stick around because it is very difficult to extract it from our sex lives. As long as it does no harm, there is no pressure to eliminate it, so it is just kind of "there."

EVERYTHING IS ABOUT SEX, EXCEPT SEX

There is a saying, often attributed to Oscar Wilde: "Everything in the world is about sex, except sex. Sex is about power." Wilde almost certainly did not say this because the use of the word "sex" to mean "sexual intercourse" did not begin until decades after his death. Nevertheless, this is a profound insight, and most psychologists will tell you that there is a complex truth behind it.

Much of what animals do revolves around sex. It occupies a great deal of their time if you consider the pains they take to get it, the time they spend doing it, the efforts they engage in to make it successful, and so on. Consider the sockeye salmon that will swim for hundreds of miles (upstream!) in order to spawn once and then die. Fish do not even get to enjoy interactive sex like we do—they just squirt their gametes. And yet, look at the lengths they go to. Look at the elaborate coloring of the golden pheasant—all of that flash and flare in order to get a mate.

For mammals, sex is much more fun, and, understandably, we spend even more time doing it and trying to do it. Even sexual activity that does not seem connected to reproduction might actually be, at least in part. For example, sex is used to establish relationships and alliances and secure one's place in the power structure of the group. Sex is used to exert and establish dominance. Sex is used to play tricks on, distract, and exhaust competitors. Sex is used to resolve disputes and maintain group cohesion. Sex is used to strengthen pair bonds and promote good family life for raising offspring, the topic of the next chapter. All of that does have an impact on reproduction, if indirectly.

If we are talking about "sex" as everything connected to mating, reproduction, and child-rearing, then yes, just about everything an animal does is about sex. With this extended definition of sex, it turns out that the "purists" might be more correct than we originally let on. We began this chapter by saying that sex is about a lot more than just making babies. What I meant was that sex is about a lot more than *conceiving* babies. If we consider pair-bonding, herd cohesion and harmony, social hierarchy, or just plain having fun, the benefits and purposes of sex are built around providing for the future of the species. It is all intertwined and cannot easily be separated. When it comes to animals in the wild, it is all about sex.

The question is, are humans any different? Remember that I am not just talking about conception, and I am not even just talking about the sex act itself. For many of us, much of our lives could be summarized as making ourselves attractive to a potential mate (physically, financially, in terms of power status in the social hierarchy, and so on), searching for and selecting that mate, establishing the resources and structure to provide for the success of our children, having children, caring for those children, preparing them for their own success in this cycle, retiring from our own family building and putting resources into the success of our children's children, and so on. These may not be our conscious motivations. Drives and urges often work subconsciously.

I am also not saying that we do not do other things along the way, but the major events and aspects of our lives involve our attempts to contribute to the next generation of our species the best that we can. Some of us are better at it than others. Others opt out of this altogether. Still, we all live in this herd together. It takes a village to raise a child, and, like it or not, we are all part of the village.

FURTHER READING

Judson, Olivia. *Dr. Tatiana's Sex Advice to All Creation: The Definitive Guide to the Evolutionary Biology of Sex*. New York: Holt Paperbacks, 2003.

Roughgarden, Joan. *Evolution's Rainbow: Diversity, Gender, and Sexuality in Nature and People*. Berkeley: University of California Press, 2009.

5

DO ANIMALS FALL IN LOVE?

N THE PREVIOUS chapter, we discussed the many uses of sex besides simply conception. However, we left out one big item: sex as means to create
and strengthen bonds. This is such a large aspect of sex that it requires its
own chapter.

As we speak exclusively about the bonding and attachment functions of
sex, it is important to keep in mind that this is in addition to the functions
that we have already discussed. In any given species at any given time, sex
is used for some subset of these many functions. For example, two hawks
might find each other while out hunting and have sex for fun and possible
procreation. Then, they might go home to their spouses and have sex to
strengthen their pair bond. These are overlapping, not conflicting, purposes.

SEX FOR PAIR BONDING

It was once thought that monogamy was a purely human affair and ani-
mals had sex with as many mates as they could. The reality is not that simple.
In birds, more than 90 percent of species exhibit "seasonal monogamy."[1]
That is to say, they hook up with a mate for one breeding season and stick

with that mate for the entire year. Birds are nest builders, and most of them do this in monogamous pairs. The roles vary among species, but in general, the birds split duties, such as foraging for nest-building materials, constructing the nest, finding food and sharing with the spouse, incubating the eggs, feeding the young, chasing off predators, and so on. It is a real partnership, and both members of the pair contribute.

For most bird species, about half of "marriages" dissolve at the end of the season when they migrate. This is because flocks of birds are not stable from year to year. Birds can switch flocks, often in search of a better position in the dominance hierarchy. You see, mate selection in many bird species follows the dominance rank order very closely. The highest-ranking females pair with the highest-ranking males and so on down the line. Rank is determined differently in each species. Size, strength, plumage, and the size, quality, and position of the nest are all things that birds may use to attract mates. The most desirable females end up with the most desirable males.

I think it is safe to say that humans are no different in this regard. The phrase "way of out of his league" comes to mind. In many birds, both males and females can attempt to climb the rank ladder by jumping flocks periodically. Humans do this, too. If we consider a season-long pairing of birds to be a "marriage," the divorce rate of 50 percent is pretty darn similar to what is seen in humans. (I mention this as a curiosity, not a mathematical correlation.)

One thing to be clear about: pair bonding in the animal world usually involves what is called economic or social monogamy, not sexual monogamy. Exclusive sexual monogamy is indeed uncommon in animals, but economic monogamy is not.[2] Economic monogamy means that two animals live together for an extended period, to the exclusion of living with others, for the purpose of tending to their young—in human terms, the making of a family unit. In some cases, these pair bonds are very short-lived. They can last just minutes, as in squirrels, or they can last an entire lifetime, as with majestic bald eagles. Lifelong economic and sexual monogamy, though rare, has been described in a variety of species, including angelfish, swans, lobsters, turtledoves, and African dwarf antelopes. In gray wolves, the alpha male and alpha female form a lifelong pair and lead their pack, which is usually made up of their siblings and children, some of which will remain celibate and help raise their younger siblings or nieces and nephews.

Economic monogamy usually involves a lot of nonprocreative sex between the members of the mated pair. Why would already-paired animals continue to have sex even when they are not fertile? One reason is that sex strengthens the pair bond. Sexual activity is known to release hormones and neurotransmitters. This time, I am not just talking about those in the pleasure–reward center. Another hormone that plays a role in sexual intimacy is oxytocin. It has been known for decades that oxytocin is released during sexual activity in many animals, and the result is the promotion of pair bonding with the sexual partner.[3] In other words, this hormone makes an animal feel "attached" to the one that they just had sex with. The oxytocin effect strengthens with repeated exposure and weakens over time if not reinforced through periodic "booster" doses.

Why would this have come about? If a male and female have sex and conceive a litter, in some species and environments, that litter will have a better chance of survival if the parents stick together and care for the young as a team. Pair bonding helps biological parents behave as parents and care for their children as a coordinated unit. If that litter has a better success rate, they will be more likely to grow up and pass on this pair-bonding behavior. Like other neurohormones, oxytocin can make individuals, even humans, feel certain things without really understanding why. One of those things is attachment to a sexual partner. For this reason, oxytocin has been called the "love hormone," though this is dangerously misleading considering some of the other behaviors that oxytocin can induce, including violence.[4]

If this hypothesis were true, we would expect to find this oxytocin effect more often in animal species that require extensive childcare, and not in species with little or no parental investment. We expect this because economic monogamy does not offer much benefit to a species that does not parent their children anyway. Indeed, this prediction appears to be true. Many species of mammals and birds show oxytocin-related pair bonding, while it has not been observed in any reptile, fish, or amphibian species.[5] These "lower" animals do have their own versions of oxytocin, and it also functions in reproduction in various ways, but it does not seem to promote monogamy. This supports the notion that this new function of oxytocin, pair bonding, evolved as a way to promote good parenting behaviors in the ancestors of birds and mammals.

We are not quite done with oxytocin. Oxytocin is also famous for being one of the hormones produced by nursing mammalian mothers, including humans.[6] Without it, nursing cannot occur, even if the breasts enlarge and produce milk due to the work of estrogen and prolactin, respectively. At the same time, we already know that oxytocin promotes bonding and attachment, so, as you can probably guess, this hormone also supports the development of the bond between a mother and her nursing child.

It does not end there, either. Oxytocin causes new human mothers to become very protective of their children and mistrustful of strangers.[7] Paranoid, even. That is really clever because right at the moment when we need our moms to provide us the most care and give us the most protection, they are strongly motivated to do so by their own brain chemistry. Although this protection/paranoia-inducing property of oxytocin has so far only been proven in humans, it seems likely that it is also behind the well-known protective nature of other mammal mothers toward their young.

The neurological mechanism of oxytocin's protectiveness-promoting properties likely contributes to the seemingly contradictory functions of oxytocin. While this hormone can produce feelings of attachment and, if we can use this word, love, for members of our family unit, it also promotes intense suspicion of those outside that unit. This helps to explain why oxytocin is also linked to xenophobia, racism, paranoia, and even violence.[8] In one hopeful study, oxytocin was given to people with borderline personality disorder to see if it would help promote the empathy and altruistic tendencies that they lacked. The result was the opposite. Since they had no bonds that mattered to them, they experienced only the dark side of oxytocin and they became even more antisocial.[9] Oxytocin is much more complicated than the misnomer of "love hormone" would imply. While it clearly promotes attachment, it can have the opposite effect toward those that are not the object of the attachment.[10]

In sum, oxytocin is a hormone with physical effects, but it also changes how we feel. The same hormone is used to promote bonding between sexual partners and to release milk, and, not surprisingly, we see that mothers and babies develop a strong bond as well. This is another case of something—in this case, a neurohormone—originally performing one function and then being co-opted to perform another, related function.

In order to do its thing, oxytocin is released in response to what we see, feel, or hear. It is released during sexual activity, even at the arousal stage. It does not make us feel sexy; it gets released when we *are* feeling sexy. Similarly, it is released in nursing mothers upon the sensation of suckling, which then allows milk to be released. It can actually begin even before that. Many nursing mothers will inadvertently begin releasing milk when they merely hear cries from a baby—any baby. When I was a baby, my mother once had to bolt out of the supermarket because she felt a wet spot on her shirt after she walked past someone else's crying baby. This oxytocin is powerful stuff.

This role of oxytocin in sexual activity and pair bonding has been reported in animals for some time, but the research in humans is the most telling. In a 2012 study, researchers found that some men who were given a dose of oxytocin would actually move away from women they did not know, but found attractive, in a social setting; but not all men. Only those who were in long-term relationships would move away.[11] The extra dose of oxytocin suddenly reminded these men that they were already bonded with someone else and that they should keep their distance from the forbidden fruit in front of them.

One final oxytocin example: this one comes from voles, small rodents that look like very small mice. These little guys are famous for being among the most monogamous, both economically and sexually, of all mammal species—certainly the most monogamous of the rodents that we know of. Researchers did an experiment with voles in which they deprived females of their natural oxytocin and observed that the preference for their mated partner was totally lost. These females would freely mate with any male—that is, any male that had not pair bonded with someone else (the males are monogamous, too). However, if given oxytocin replacement with injections directly into the brain, the females would regain their monogamy and once again prefer only their mated partner.[12]

This observation with voles and the one above with the men who avoided attractive women prompted some therapists to study whether oxytocin could aid couples in marriage counseling. In one oft-cited study, it seemed to work as hoped.[13] However, in another, it promoted intimate partner violence.[14] Oxytocin is complicated and its therapeutic potential in relationships must be explored with great care.[15] Just as in the treatment of chronic mental illness, brain chemistry is intensely complicated and often resists our ability to manipulate it pharmacologically.

Nevertheless, given the parallels in oxytocin effects between humans and other animals, humans really do seem hardwired for pair bonding, and sex promotes that type of attachment. This will come as no surprise to psychologists. But what does it really say about human monogamy and marriage? As you can probably guess, my opinion is that it tells us that we are not that different from animals in this regard, just more emotionally sophisticated. (Some of us anyway.)

The consequences of hormone-mediated attachments are not always positive. In addition to the tendency to promote distrust of nonaffiliates, hormone-mediated attachment can maintain an unhealthy pair bond. I bet most of you know a woman or man who has stayed in a relationship that everyone else knows is bad for him or her. Why do people do this? Obviously there are very complex reasons stemming from underlying psychological issues, but I am not the first to suggest that biology may also be to blame. If we know that oxytocin is hard at work promoting attachment and that sexual activity also stimulates the release of dopamine and serotonin in the pleasure centers of the brain, we end up with a powerful chemical cocktail that drives a person to stay with the "bad guy" (or girl). It is very much like being addicted to a drug, involving many of the same neurohormones and brain centers. Partly for this reason, a bad relationship can be a hard habit to break.

INFIDELITY? THE CURIOUS CASE
OF THE CLIFF SWALLOWS

As I have already mentioned, many species of birds that are impressively faithful in their economic monogamy do not always employ sexual monogamy. Is this like cheating? The answer appears to be "no," because the behavior is often not hidden, is tolerated in the marriage, and is engaged in by both parties. It is not cheating if both parties are OK with it. Even in humans, some people choose to be in open marriages or reach some sort of an understanding about extramarital sexual activities. These arrangements might not be for all of us, and some can argue about the morality or propriety of them, but they do seem to work for some people. They certainly work for most birds.

The cliff swallow is a good species with which to highlight the complexity of avian social structures and the way that economic monogamy is totally separate from sexual monogamy.[16] These birds live in very large colonies that include thousands of nests that look like large tubes made of dried mud stuck to the side of a cliff. Nests are always built and tended to by bonded pairs. For a cliff swallow, part of one's status in the colony depends on nest location. The most desirable spots are near the bottom of the cliff nearest to the "public areas" in the center, presumably because they have the easiest access to the water source for bathing and drinking and for interacting with other birds. After all, it is fashionable to be seen in the hot spots, right? The less desirable locations are on the outskirts, and the lowest-ranking birds in the colonies reside in separate sub-communities, like bird suburbia.

These birds have very active social lives. Pairs work together to build and tend to their nests, sometimes stealing materials from other nests for their own, and so on. A pair may even occasionally attempt to take over another pair's nest in a more desirable location. However, much of the socialization revolves around sex. For example, when females leave their male partners to guard the nest while they forage for more mud, they often engage in copulation with other males while they are at it. This is called an extra-pair copulation (EPC). In other words, while they are out getting groceries, the cliff swallow wives go ahead and get a little something on the side. How do their husbands respond to this? They have sex with them the minute they return.[17]

Why might this be? This is sometimes characterized as "monogamy betrayal" on behalf of the female.[18] In response, the males have sex as quickly as possible in an attempt to remove the enemy sperm and replace it with their own. In other words, this hurried sexual event is nothing more than a male's attempt to protect his paternity of future children.

There are other interpretations as well, spurred by a few complicating observations.[19] First, the cliff swallow pair will have sex when the female returns, whether she is currently fertile or not. Second, the sperm-replacement technique does not work in this species. (It does in some others, however.) Third, these "welcome home" copulations occur with equal frequency when the female has engaged in an EPC and when she has not. Fourth, the female does not resist the spousal copulation. Resistance to copulation is observable in cliff swallows, and it is not typically seen in these instances. Fifth, as already alluded, the EPCs are done in full view of the colony; there is no apparent effort to be clandestine with sexual activity outside the pair bond.

A simpler explanation may be that these cliff swallow pairs simply "miss" each other when one is away from the nest doing chores. I find this to be a simpler explanation because the two birds are pair bonded and we know that attachment drives individuals to be with each other socially and sexually. When members of a bonded pair are separated, they "yearn" for each other and celebrate when reunified. Sex is a product of, and reinforcement for, pair bonding in part because oxytocin, released during sex, reinforces the attachment.

Rather than an attempt to fight back against cuckoldry, these "welcome home" copulations may be just that—a way that these two pair-bonded birds greet each other after some time apart. I suspect that their concept of time is not as developed as ours, so a quick errand to get some mud might feel like weeks apart. This hypothesis still needs experimental evidence, of course, but I think it is a good place to start.

But what about the EPCs themselves? Because they are not hidden, it may be that EPCs are a normal and accepted practice in cliff swallows. Considering that they happen in broad daylight, in the most public spot of the colony, and in the full view of everyone, including spouses, this does seem like the most likely scenario. Cliff swallows are swingers. So what?

There are some interesting consequences for this. First of all, every so often, we should expect extra-pair paternity (EPP), in which the male of a mated pair spends his time and resources helping to raise another male's offspring. Indeed, that is exactly what is seen. Using genetic testing, scientists observed that EPP existed in 24 percent of the cliff swallow nests that were tested in one study.[20] That is actually somewhat lower than was expected, given how much hanky-panky is going on with these birds.

Viewing this through the cold Darwinian view of constant competition, our hearts go out to the poor cuckold, and we see him as a hapless victim. As usual, the real story is deeper than this. It turns out that when researchers tested these "poor cuckolds," they discovered that they were the most active of the male extra-pair copulators themselves.[21] In other words, "cheating" males tend to end up with cheating females.

Extramarital sex is not the only shenanigans in these cliff swallow colonies. It turns out that these birds, usually the females, will sometimes place their eggs in the nest of other pairs.[22] In this scenario, these "parasitic" mothers are actually pushing their own genetic offspring on to other birds. This allows them to have even more children than normal, exploiting the parenting effort of other bird pairs. This is referred to as nest or brood parasitism. It has not yet been tested whether these egg-passing events are actually the actions of "cheating" mothers returning eggs to their true father, but that seems unlikely. Just like with the "cheaters," the tendency toward "egg passing" was stable in an individual over her lifetime. Some birds are just inclined to put their eggs in other birds' baskets.

For some of the researchers who have spent so much time cataloguing the interesting behavior of these swallows, these birds are nothing but a bunch of liars, cheats, thieves, and cuckolds. Others suggest a different view.[23] Maybe these swallows have a system of distributed parenting and paternity. Perhaps each pair and its young are not the family unit in these birds, but maybe they are more like one big family. Maybe the "pairs" are merely the unit for raising children, not necessarily making them, because pairs have been proven to be the most effective or efficient at incubating the eggs and raising the young fledglings. Perhaps these large colonies are more like a pride of lions or a wolf pack. The difference is that instead of true group parenting, like in lions and wolves, these animals break down into pairs for the purpose of egg incubation and care of the hatchlings.

I find this alternative especially appealing considering the fact that extra-pair copulations may be a key triggering event in the evolution of colonial living in birds.[24] This theory holds that at least one incentive that a male or female bird has to live in close proximity to other birds is to pursue EPCs. While the cuckold might have a reason to try to stop EPCs by his/her spouse, the cheaters themselves stand to gain substantially by sharing their genetic potential with as many as possible. Colony living makes that easier. While this inevitably sets up a "battle of the sexes" style conflict, communal paternity may be one way that birds have evolved to mitigate that struggle.

In addition, the more traditional view that EPCs are purely deceitful, parasitic, and ruthlessly competitive, requires some assumptions that I find problematic: These poor birds cannot spot their spouses cheating on them in plain sight, despite having excellent vision. They cannot stop others from placing eggs in their nest while away from the nest for mere minutes. They cannot spot a fresh, warm, foreign egg on top of their own eggs laid days before. Some of the birds are completely helpless when it comes to thwarting this deception and exploitation, while others are ingenious at executing it. In summary, you would have to believe that these birds are both fiercely competitive and very stupid. Keep in mind that these same birds have learned how to build nests on the side of a stone cliff without the use of fingers or hands. And yet they are hopelessly dim?

On the other hand, the view that cliff swallow flocks operate as one big family has some complications as well. Most importantly, it is difficult to imagine how cooperative parenting could emerge when an individual has such a powerful incentive to "cheat." Prosocial behaviors would not be rewarded by natural selection and selfish ones would be. Remembering that selection generally operates on the level of the individual, maintaining such distributed cooperation would be hopeless when individuals have no incentive to pull their weight.

However, there is one way in which the cooperative parenting model does favor the interests of individuals and the particular genes that make it possible. This is referred to as kin selection, when the group members in question are all close relatives.[25] Kin selection can operate to promote close cooperation in groups of individuals that are closely related because the genes that promote the cooperation also promote their own successful reproduction. Under the right conditions, this can be a powerful force and helps explain the social behaviors of a diverse array of animals from ant colonies and beehives to wolf packs and prides of lions. Genes that lead you to help your close relatives are self-promoting, provided that the relatives share those same genes.

Could kin selection be at work in cliff swallow colonies? Possibly. Studies have shown that they are quite inbred. Each member of the colony has a 50 percent chance of being a first- or second-degree relative to each other member.[26] While this is not as closely related as members of a wolf pack or bee hive, it is at least possible that this is related enough to promote kin selection-mediated emergence of cooperation in child-rearing. I find this even more appealing when you consider that the cliff swallows do have quite a bit of competition as well. It is not all cohesion and harmony. As I mentioned before, pairs of cliff swallows attempt to evict and steal nests from better-placed rivals. Some pair bonds are dissolved because one member finds a better mate. The lower degree of relatedness than that found in some other kin-selected social groups predicts a lower degree of cooperation and a higher degree of competitiveness.

When you factor in the EPCs and the brood parasitism, a staggering 47 percent of all cliff swallow nests produce at least one fledgling per season that is not the biological offspring of one of the parents.[27] Of course, this could be because of the fierce competition that goes on, but I support the opposite view: that distributed and cooperative parenting is part of the social fabric of the species, even if it is associated with some tension and healthy competition.

Another thing that would be predicted by the more "progressive" view of cliff swallow extended family life is that copulation is about much more than just paternity and conception. I have already noted that it promotes bonding between the married couples. What I failed to mention is that the mud hole also sees a lot of male-on-male action.

Because these birds have no external genitalia or sex-specific markings, researchers have a tough time quantitating the male-male sexual activity by simply watching from afar. It is easy to tell when a male bird mounts another bird because females do not seem to mount at all, but it is not easy to know for sure if the mounted bird is a female or another male. Instead, the researchers used taxidermic stuffed swallows and observed that 70 percent of the copulation events were males mounting the stuffed male rather than the stuffed female.[28]

More on same-sex sexuality in a moment. For now, my point in telling you this gigantic story about swallows is to highlight how issues of sex and family are complex—not just for humans, but for all animals. When it comes to observing things like sexual activity, pair bonding, fidelity, and parenting, there is usually much more to it than meets the eye. We all know that you cannot really learn much about a married couple by observing them casually at a distance for a few hours. The same is true for animal species. A scientist must keep a watchful eye and a mind open to possibilities that do not conform to the sexual norms she is accustomed to in today's society. As with the cliff swallows, sex might be used as often to promote pair bonding and colony cohesion as it is for procreation.

Another reason that I have spent so long discussing these swallows is to introduce a theme that has come up time and time again in this book and will continue to: cooperation. Many people, even some biologists, regard natural selection as a cutthroat competition between fellow members of the same species. The many struggles of life are boiled down to the phrase "survival of the fittest." This can be misleading, especially because we often forget that "fitness" is measured not by strength, health, or vitality, but in contribution of one's genes to the next generation. Many animals, particularly birds and mammals, have long figured out that cooperation is just as good a path to fitness as is competition.

Cooperation is particularly beneficial for the process of natural selection when the cooperators are closely related to each other. In this case, the relatives share many genes already, including, presumably, the genes for cooperation. One can easily imagine a scenario when "cooperation genes" came into existence through mutation and were spread to offspring for a few generations. If these cooperation-prone animals then work together and breed together, they may be more likely to survive and be successful. For this to work, the communities must be relatively small, close-knit, and inbred. Over time, the cooperation genes spread and eventually take over the population as the individuals that cooperate fare better than those that do not. Importantly, even an individual that does not breed can still promote the persistence of the cooperation genes by helping out siblings, parents, and cousins. By helping your close family, you are promoting your own genetic stock.

This is not to say that within cooperating communities there is no competition. There certainly is. Each pair of cliff swallows is always vying for the best nest, and fights over nest location do occur. There are sometimes even evictions and attempted evictions as the dominance hierarchy plays out each season. After all, a little healthy competition is good for a breeding population, even when the population is marked by substantial cooperation.

In fact, there may even be extra pressure to "weed out" freeloaders in a cooperative society. In a purely competitive social structure, any lazy or otherwise suboptimal individuals would quickly perish under the weight of their own problems. In a society of shared labor, however, these pure "takers" may attempt to slide by undetected, benefiting from the cooperation of others while donating little effort in return. Thus, intra-flock competition helps keep everyone on their toes and performing at their best.

In summary, the life of the cliff swallows, with all of its egg-passing, extra-pair copulations, sex-heavy married life, and shared parenting, is best seen as a mix of cooperation and competition. It is more like a basketball team than a track-and-field event. Each member may try to "stand out," but the good of the team comes first.

You may wonder why I have spent so long discussing this single species of bird. I have done so not because cliff swallows are special, but because they are not. We know so much about this species only because scientists have taken the time to watch them. Charles and Mary Brown have been studying one particular colony of cliff swallows for more than thirty years.[29] The story seemed so simple at first, but closer inspection revealed a staggering complexity and nuance in their social dynamics. It seems rather likely to me that just about every species has similar complexity and nuance, just waiting to be discovered. I wonder how many species we have misunderstood because we have not taken the time to observe and study them carefully.

A WORD ABOUT MONOGAMY IN HUMANS

In animals, we saw that sexual monogamy is pretty rare. Instead, social/economic monogamy, without complete sexual fidelity, is quite common. Ninety percent of bird species are monogamous in this way and about half of mammals are, with most of the remainder engaging in variations of community-based family (prides, packs, and so on) that are essentially extensions of economic monogamy in terms of mutual investment in "marriage and family life."

In my view, economic monogamy is the better parallel to human marriage than sexual monogamy. Hear me out. I am not saying that most human married couples are not faithful; I am not saying that we should not even bother trying to be faithful; and I am not saying that sexual fidelity is not an ideal that we should strive for. What I am saying is that we should be honest with ourselves about this topic. First of all, we know that many marriages do not enjoy perfect lifelong sexual monogamy. I might even say that most marriages eventually struggle with fidelity and temptation in some way. In fact, does anyone disagree that lifelong marital faithfulness requires constant effort and ever-renewing commitment? It might even be said that being faithful feels to many, at one time or another, contrary to their natural urges and appetites. Any priest/minister/counselor/friend will tell a couple intending to marry that they may have struggles ahead in this department. This is just a fact of life for us.

To admit this truth is not the same as making a biological argument for the abandonment of marriage or faithful monogamy. That is not my intention. Far from it—I am a happily married man in an arrangement that we have both decided will be sexually monogamous. Personally, I accept that I may be "going against my nature" a little bit in order to embrace monogamy. I am happy about that choice, and I make it freely. After all, there are many behaviors commonly seen in nature that human beings have decided are no longer fitting for us, such as murder and infanticide (see chapter 7 on jealousy).

Furthermore, it has long been known that restraining some of our appetites and urges can be in our best interests (see discussion of gluttony in chapter 8). In fact, in my view, knowing the biological reality that sexual monogamy can sometimes be a struggle offers a more honest and realistic path toward achieving it. It facilitates candid discussions about sexual needs, reduces pointless guilt about temptation or near misses, and allows understanding and reconciliation should these be required.

SAME-SEX SEXUAL ACTIVITY

For many biologists, homosexuality was previously seen as a conundrum. First, they did not think that animals experienced any same-sex attraction or purposefully engaged in homosexual acts. So where did it come from in humans? Second, homosexuality, in humans or other animals, seemed contrary to what natural selection would predict. It would seem that same-sex attraction would tend to reduce fertility. If a trait reduces fertility, even a little bit, it will quickly be eliminated, as the bearers of that trait would leave fewer offspring—right? For these reasons, many people considered homosexuality to be unnatural, an aberration seen only in humans and thus probably psychological, not biological. I will not go into all the harm that this kind of thinking has caused, but suffice it to say that for a long time, many people—even scientists—considered same-sex sexual activity to be unnatural.

Once again, it turns out that we were wrong. Both of these claims were completely inaccurate. First, animals most certainly do engage in same-sex sexual activity. A lot. In every species in which it has been looked for, it has been found. As of 2008, no fewer than 1,500 animal species have been observed to engage in sexual activity with members of the same sex.[30]

Also, it turns out that same-sex sexual activity does not seem to reduce reproductive success, either in animals or in humans. Although the data are still coming in, several possible explanations have been proposed. First, we should discuss the prevalence of same-sex sexual behavior in animals. I begin with selected examples from humans' more distantly related creatures and build up to our close relatives. I should also say that I am well aware that applying the term "gay" to animals is not exactly accepted terminology. I do not advocate that it should be. I am just using it to lighten things up. The term "same-sex sexual activity" is pretty clunky to write and boring to read. Please forgive me the impropriety if I take a shortcut here and there.

Bedbugs have been found to engage in male-on-male sex. I am not kidding—gay bedbugs! This example is kind of a letdown because it seems to be nothing more than a case of mistaken identity. A male bedbug will usually mount and inseminate a female bug that has recently fed. Sometimes, however, he will mount and inseminate a male that has recently fed. Insemination in bedbugs means using a needle-like organ to pierce the body and inject semen directly into the abdomen, a phenomenon called traumatic insemination, which can cause injury, infection, and death.[31]

Not surprisingly, male bedbugs will resist being mounted, given how injurious it is. Sex injures females also, but unfortunately, that is just how it goes for bedbugs. These injuries are the cost of reproduction for female bedbugs. When a male bedbug injects another male, however, there are no winners. In fact, it was recently discovered that male bedbugs will emit an "alarm" pheromone after eating that attempts to signal to other males that they are male and that insemination would be pointless.[32] That this pheromone is not 100 percent effective argues that this antirape defense could be a relatively new evolutionary innovation and the species is still growing into it.

I offer bedbugs as an example of an invertebrate (an animal with no backbone) in which homosexual activity has been described in some detail. Other invertebrates with a pension for gay sex include blowflies, wasps, fleas, silkworms, octopus, flukes, locusts, and butterflies. Until recently, most biologists chalked all of this same-sex sexual activity up to mistaken identity. That may be a likely explanation for bedbugs, given that they have poor eyesight and their sense of smell is focused pretty exclusively on the detection of blood, whether as food or in the search for a well-fed female. However, mistaken identity is a pretty poor explanation in most other species. Many invertebrates have incredibly keen eyesight and smelling. I find it exceedingly unlikely that insects that can easily tell a virgin female from a nonvirgin would be totally helpless in distinguishing males from females.

Same-sex sexual activity has also been noted in all the major divisions of vertebrates: the older ones (fish, amphibians, and reptiles) and the newer ones (birds and mammals). In chapter 4 we talked about some of the courtship behaviors of multiple-gendered male fish and the "female mimicry" of some Canadian garter snakes that trick the males into having sex with them in order to steal body warmth.

Turning our attention to birds, same-sex sexual activity has been reported anecdotally in zoos and sanctuaries since time immemorial. In the past, this forced some uncomfortable acknowledgement from biologists that same-sex sexual activity had occurred, but it was usually characterized as a strange side effect of captivity or small population size. This is sometimes called the prison phenomenon, and I think you can imagine why. To me, the fact that it was called the prison phenomenon exemplifies the prejudice that many scientists held for so long. They refused to believe what was happening right in front of their eyes. Because of this, the long-held view was that same-sex sexual behavior in animals was rare, an artifact of captivity, and occurred only when access to the opposite sex was limited or impossible.

Farmers have known better all along. While most bulls are aggressive toward other males and are usually kept in a separate paddock to avoid fights, there is occasionally a bull that prefers to mount other males. This mounting can lead to ejaculation if the mounted bull allows it, and he sometimes does. Farmers can usually get these male-preferring males to impregnate females by simply denying them access to males. (I bet no one calls that the prison phenomenon.) But even that method fails sometimes: some bulls simply will not voluntarily have sex with a female cow.[33]

This reminds me of what happens with human gay males. Most of them will admit to having had sex with a woman or two before coming out of the closet, while a few remain "pure." There is even a term for this: a goldstar gay is a man that never once engaged in any sexual activity with a woman. I digress. Actually, no. This a valid comparison, is it not? Most of the gay bulls can perform with a female if they really have no other choice, and the same is usually true for gay men. For most of the last few centuries, they really did have no other choice and so most of them remained in the closet, married, and fathered children.

Anyway, seemingly without exception, all of the herd-living mammals show plenty of male-on-male mounting, penetration, and orgasm. Bison, wolves, gazelles, antelope—you name it. The kob is an animal that looks like a cross between an impala and an antelope and lives abundantly in the African savannah. It turns out that, when population size allows it, the females will form large female-only herds and engage in lots of sexual activity among themselves, including genital licking and rubbing, mounting, and general affection, such as licking-kissing and rubbing heads. They will still get pregnant by allowing an occasional male into the fun, but the male does not stick around after the sex, nor do the females seem to "enjoy" the procreative sex when it does happen.[34]

One of the gayest species of all is the giraffe. One study found that 94 percent of the time that a male giraffe mounted, he was mounting another male.[35] Ninety-four percent! Other studies have put the estimate of same-sex mountings between 30 and 75 percent. Either way, that is a lot of same-sex sexual activity. Some biologists of the older and more conservative persuasion get rankled if we refer to these mountings as sex. They would say that this mounting behavior in giraffes and other animals is about dominance, not reproduction, and thus, it does not count as sex. Huh? Only reproductive sex actually counts as sex? Really? So then, if a woman has sex with her husband while she is on the pill, does that mean that they did not actually have sex? By this bizarre definition, two men or two women can never actually have sex, which would come as quite a surprise to the LGBT community.

Sex is sex, and the refusal to see nonprocreative sex as true sex brings us back to the deep-seated prejudice and scientific bias with which we began this discussion. The fact that male giraffes have sex with each other for reasons that are obviously not procreative is precisely the point. Sex, in animals and humans, is about more than just procreation.

In the case of these male giraffes, they mount and penetrate each other to the point of climax, meaning that, yes, they have orgasms when they do this. What I think is even more interesting is that they will do this after a battle for dominance. Not before—after.[36] One giraffe has already won, and the other has yielded, and only then do they "get busy." Scientists that want this to be purely a dominance display are really grasping at straws because the struggle for dominance is over by this time. Furthermore, the sex that the males have is not violent, and the vanquished male is not forced into submission as if it were rape. The giraffes engage in "necking," which is just what it sounds like—mutual neck rubbing. They are affectionate and playful for a while before the dominant male mounts. That sure sounds like foreplay to me. The interesting question is, why? Why do they have sex after the dominance struggle is over?

For giraffes, scientists do not have solid answers on that yet, just descriptions of the behavior. In many other species, however, there is a much fuller understanding of the many functions and purposes of same-sex sex. Some of these are listed on the pages ahead, and maybe one of them applies to these gay giraffes.

The rampant bisexuality and homosexuality of dolphins has been well reported in the scientific literature and the popular press. In one study of

120 male bottlenose dolphins, researchers found that they not only spend almost all of their time with other male dolphins, but they also spend a lot of time having sex with them.[37] In fact, dolphins often travel in groups comprised entirely of one sex. The all-male groups will engage in sexual activity quite frequently, including body contact that is genital-anal, genital-genital, and oral-genital. (They will even penetrate each other's blowholes, giving new meaning to the term "blow job.") This body contact between males includes orgasm/ejaculation and often involves more than two males at a time. Orgies of up to fourteen males have been reported.

While female dolphins are not as sexually active with each other, female-female sex has certainly been seen as well.[38] When it comes time to actually breed, a group of males will have to work together to herd and corner a group of females. The females do not appear to go willingly, but—as is often true with animals—it is difficult to know for sure if the chasing, resistance, and forced sexual contact are truly violent or just part of the expected courtship. After all, some like it rough. One thing is certain, however. After the brief, procreative coed encounter, both the males and females are eager to get back to their sex-segregated pods and all the orgies that await them.

Rams (male sheep) turn out to be the species that has grabbed the most headlines for their gayness. This is due to the discovery of slight brain differences in the gay rams compared with the straight ones.[39] This announcement was met with simultaneous applause and outrage. While some people were delighted to see hard, measurable proof that homosexuality was an innate biological feature (and thus not a choice), others feared for what might be done with this information. Would it somehow be used in the future to try to cure, abort, or breed out the gays from the human population? Some of these fears intensified when the researchers proposed to probe further so that gay rams could be identified during the process of selecting rams for breeding. They also planned an experiment to see if hormone treatment during pregnancy would alter the same-sex attraction. Much of this was exaggerated in the press, and the researchers were demonized by great mobs because of erroneous reports about what they actually intended to do.

Controversy aside, the discovery itself is fascinating. Working with a population of domestic sheep, researchers put single rams together with pairs of restrained ewes or rams and observed their mounting behaviors. They found that 7 percent to 10 percent of the rams would not mount ewes at all but would aggressively mount the restrained rams and ejaculate. Those are the "gay sheep" that you may have read about, but the study found more than that. About 20 percent of the rams would mount both ewes and rams, reaching orgasm with both. I suppose these would be bisexual rams. A whopping 15 percent would mate with neither. Can we call them asexual rams? This study has since been repeated several times, by the same group and others, with slight variations. A later study reported slight differences between gay and straight rams in their brain anatomy and sex hormone levels. The fascinating picture that emerges from the study of male sheep sexuality is that only a slim majority of rams are the expected male-attracted-only-to-females.[40]

Personally, I suspect that the sheep studies actually underestimate the diversity of sexual preference because they examine only anal penetration, and only from the perspective of the mounter, not the mountee. Using common gay slang, this study was only on the lookout for "tops," the ones that enjoy doing the penetrating. But what about the "bottoms?" (Bottoms are males that prefer to be penetrated, rather than penetrate someone else.) It is conceivable that at least some of the 15 percent of the sheep that were reported as "asexual" do not mount the males or the females simply because they are just not into mounting. Maybe they prefer to be mounted? Or maybe they are not into anal or vaginal sex at all and prefer oral sex or some form of frottage? Before you scoff, remember that so, too, did most people scoff at the notion of gay animals just a few decades ago.

The example of gay rams is useful because it shows how much more diversity exists in sexual orientation than biologists previously thought. In fact, sheep sexuality bears a striking resemblance to the Kinsey scale for humans, in which strict heterosexuality and strict homosexuality are at the extremes, plenty of individuals fall in the middle somewhere, and some individuals (so-called asexuals) have almost undetectably low sex drives. More study is needed, not least of all because many studies, for some reason, have only examined males. The most important lesson to take from all of these reports of same-sex sexual activity is that it has been found where it was least expected to be—everywhere.

How was all of this gay sex missed by biologists and naturalists for so long? Prejudice seems the likely reason. I do not necessarily mean that all biologists were sexist or homophobic, although many surely have been. What I mean is that if you are not looking for something, you usually will not find it. While critiquing a field biological study published in a top journal, Bruce Bagemil stated it this way: "Every male that sniffed a female was reported as sex, while anal intercourse with orgasm between males was [reported as] only dominance, competition, or greeting."[41] In other words, biologists did not really think of two males having sex as two males having sex. It had to be something else.

To be fair, observing, cataloguing, and analyzing animal behavior is a long and arduous process, so it is very easy to overlook things that you were not looking for in the first place. It is not that these biologists were blind or sloppy. They were just testing other hypotheses. Also, while pretty much all animals engage in same-sex sexual activity, the vast majority of them do not do so at the total exclusion of opposite-sex sexual activity. For most animals, there is plenty of sex with members of both sexes.

SAME-SEX PAIR BONDING

The previous section was about animals having same-sex sex, per se. Just the sex act itself. However, it turns out that the "homosexual" nature of animals is not limited just to having sex but also includes choosing "mates" as pair bonds. Our story begins with the most prolific of all pair bonders: birds.

Albatrosses are large sea birds that have distinctive orange beaks with downward-pointing hooks at the end of them. Unlike gulls, skimmers, and egrets, they have webbed feet that look like those of ducks and pelicans. They are majestic creatures that are both beautiful and ugly at the same time. As you may remember from your high school British literature class, "The Rime of the Ancient Mariner" centers on the pointless killing of an albatross. They are nearly as famous for a groundbreaking study documenting extensive same-sex pair bonding.[42]

Albatrosses live in large, stable colonies or flocks but engage in some immigration between flocks to reduce inbreeding. They generally pair bond for life, following a very elaborate dance-based courtship. In fact, these long-lived birds reach sexual maturity in five years, but most will not breed or form a pair bond for another two years, sometimes longer. It takes an average of two mating seasons for an adolescent to learn the mating dance through observation, mimicry, and trial and error. Both sexes engage in the courtship behavior, which is primarily by display only and not directly competitive or hostile. Once a pair is formed, they mate for life, and they never dance again. (Some human married couples may feel that pain.)

Each breeding season, the paired albatrosses work together to build a nest, defend it, incubate an egg or two, hunt and forage, and share food with each other and the hatchlings. There are no sex or gender roles (except in the actual laying of the egg), and duties are always shared.

In the now-famous study, hundreds of Laysan albatrosses were studied over a three-year period in Oahu, Hawaii. Of the 125 nests that were built by bird pairs, thirty-nine of them were tended by a female-female pair. That is one-third of the nests, representing half of all of the pair-bonded females. (Colonies have lots of unpaired birds of both sexes that are either widows looking to pair again or those that have not yet learned the dance or attracted a mate.) It is true that there were fewer males in this particular colony (41 percent), but that does not come close to explaining the high rate of female-female pairing. This is no "prison phenomenon," and we are not just talking about sexual acts that some stubborn biologist will insist on explaining as something else. One-half of the pair-bonded females were lesbians—they selected a fellow female bird on purpose and stuck with her for life.

These lesbian pairs were almost as successful as their heterosexual counterparts at producing offspring. This indicates that the birds were engaging in EPCs, as we saw with cliff swallows. The two females engaged in alternating maternity such that nearly every incubated egg belonged to one of the pair. (The occasional "offspring" that belonged to neither was the result of other birds putting eggs in their nest in a rarely successful but frequent attempt to distribute offspring and parasitize parenting, also as in cliff swallows.) These lesbian albatrosses are "fit" in the evolutionary sense (they leave successful offspring), and their pair bonding is just as lifelong as the opposite-sex pairs.

Prior to this study, it was assumed that a same-sex couple raising children was a rarity in nature, occurring only when the couple were siblings or otherwise closely related. Thus, the authors of this study conducted genetic tests and were surprised to see that the same-sex albatross couples had the same low incidence of relatedness that the opposite-sex pairs had. Also, the pairing of the two females occurred after the same elaborate courtship dance and prior years of trying. The pairing was not sporadic or desperate— there were still plenty of unpaired males around. These birds knew what they were doing.

It is also worth noting that these lesbian albatross pairs are just as sexually active as opposite-sex pairs are. Sure, they occasionally go off to get pregnant by males they are not bonded to, but to be fair, so do "straight" females. Why do these birds have sex with each other when the sex cannot possibly be procreative? For the same reason that opposite-sex human couples have sex even when they are practicing birth control: because they enjoy it and because they love each other. I cannot say for sure whether these albatrosses feel love for each other like we do, but we do know that they stay paired until one of them dies. This is, yet again, a demonstration that sex promotes pair bonding among animals, even when it is not procreative.

This study shocked the scientific community and was widely reported by the popular press. Gay rights groups were elated, and conservative preachers were indignant. But why all the uproar? Surely we already knew this about birds. Once again, it turned that we had seen all of this before. We just did not believe it.

In 1998, two male penguins in the Central Park Zoo in New York, Roy and Silo, courted each other, pair bonded, and built a nest together.[43] They even appeared to attempt the hatching of an egg-shaped rock. They actively tried to steal eggs from other nests. I suspect they would have been successful had they been in the wild, rather than a zoo, where the population is more concentrated and food is brought right to the nest-defenders by zookeepers. Noticing that the pair was male, the handlers provided them with a fake egg that looked and felt real. They promptly began to incubate the egg—and diligently. They took turns and primped the nest in preparation for the child. Clearly stressed when the egg never hatched, Roy and Silo got their wish when the zookeepers finally provided them with a real penguin egg to care for.

These two gay penguins incubated their egg dutifully, and a healthy baby female, Tango, was hatched. In an interesting twist, the egg that was given to them was taken from a male-female pair that had failed to hatch their eggs despite two tries. This straight couple just was not attentive enough to the incubation, and, without the proper warmth, bird eggs do not develop. This is a very good parallel to the current trend of gay human couples serving as foster and adoptive parents for children whose biological parents cannot provide the care that is needed. Little Tango was raised healthy and happy by her foster fathers. In fact, this case has served as a model for gay foster parenting of penguin eggs and chicks that has been replicated at zoos around the world.

Gay bird pairings are not just an artifact of captivity. In Australian black swans, male-male pairings are almost as common as female-female pairings are in albatrosses.[44] In fact, studies have revealed that the gay black swans are more reproductively successful than their straight counterparts. How can a gay male couple have more biological children than a straight couple? Well, these gay black swans occasionally let a female into their roost for the purpose of fertilizing and laying eggs. They take turns mating with her, and paternity is shared more or less equally. Then, the three of them stay together through the egg incubation and hatchling stage. That these eggs are nurtured by a threesome explains why they are more successful than "traditional" male-female couples. When it comes to defending a nest, incubating eggs, and protecting young hatchlings, three beaks are better than two, at least in black swans.

This is a good opportunity to draw the distinction between "having lots of children" and "reproductive success." Even if he sires thousands of offspring, an animal is not reproductively successful if those offspring do not live to sexual maturity and sire offspring themselves. Parental investment has evolved in many animals, especially birds and mammals, as a strategy to have fewer offspring and to care for each of them better. In such a strategy, pair bonding is about more than just having offspring. It is also about caring for those offspring. Among birds, pairs are usually better than single parents because they are able to divide the labor. One parent can go out to find food, while the other incubates the eggs or protects the babies. The hunter-forager must find twice as much food, but she can take all the time she needs knowing that the eggs or babies are safe. For a single parent, there is just no way to hunt and incubate the eggs at the same time.

Lest you think that I am making an argument that human single parents are not as good as human coupled parents, be assured that I am not saying any such thing. In the case of humans, we have civilization, technology, currency, society, culture, and an endless list of things that make us different from animals living in the wild, especially when it comes to meeting the needs of our children. Furthermore, I was discussing birds in this example. In most mammal species, the rearing of children is not the work of either single parents or couples—communities are involved. That is the better analogy for humans—we raise our children in a community, a society working together.

If you do not believe me, consider things like day care, playdates, babysitters, and schooling. We are constantly bringing our children together with the children of others and arranging some form of cooperative care, sometimes with an economic exchange and sometimes not. It is really not so different from when a lioness leaves her cubs to be nursed by another female in the pride while she goes off to hunt for the group. The fact that black swan threesomes are more successful parents than black swan twosomes is helpful to understand the survival pressures of their particular habitat. That habitat bears no resemblance to the modern human experience, so trying to project social values in either direction is a little silly.

My point here is to say that, if "family" is defined as individuals living together and sharing a common purpose in raising offspring, birds and mammals have long known that there are many ways to make a family, something humans are only now begrudgingly figuring out.

If I wanted to be antagonistic here, I could point out that the traditional family unit of mother-father-children is the aberration, even for humans. Before the dawn of civilization, all evidence points toward communal living, not unlike the troops of chimpanzees and gorillas that we see in Africa today.[45] The family unit was a large group in which children were raised together, nursed together, and learned together. In fact, in many developing countries, families still exist in multigenerational units with grandparents and cousins all living together. There is a very solid basis to say that this extended family motif may be the most "natural" arrangement for a human family. It is curious that few people would argue that we should all return to that.

Hillary Rodham Clinton took some flak from conservatives for her 1996 book *It Takes a Village*, which took its name from the old proverb that children are not raised by their families in isolation but within a community. Various features of the village will profoundly affect the child, regardless of the behavior or wishes of the parents, so goes the saying. Things like education, health care, public safety, mass media, and entertainment programming are just a few of the features that come to mind.

Ignoring the politics of the issues here, this is a pretty good way of thinking about how community interactions are an integral part of how humans engage in communal parenting. The actions of the community can affect the success of your children, no matter what you do. No amount of parental investment, care, and love would have prevented the early deaths of most children born in the plague-infested days of the twelfth century. Similarly, even the most negligent parents are unlikely to lose a child to polio or smallpox these days. Working together, we have taken measures to ensure the survival and success of our children.

Human society, government, schools, and other institutions are merely our ways of formalizing the incredible amount of mutual investment we all have in each other, especially in our collective children. What is good for all of us is good for each of us, and vice versa. At the level of biology, we are all interconnected as one extended family, just like those cliff swallows, albatrosses, penguins, and giraffes.

BIOLOGICAL BENEFITS OF HOMOSEXUALITY?

I have already mentioned a couple of scenarios in which homosexuality actually provided a fitness benefit. In the black swans of Australia, two pair-bonded males entice a female surrogate to make a temporary arrangement for baby-making. In penguins and many other species of birds, two males can raise youngsters with alternating paternity without having to lay the egg. Egg laying, while not as demanding as giving birth to live young, still requires a substantial investment from the energy and nutritive resources of the mother.

Those are simple examples, but more interesting ones exist as well. For example, in any social species with sex-specific dominance hierarchies, there would be a clear benefit for an individual to form a pair bond or close alliance with the highest-ranking individual that he or she can, regardless of their sex. There would be a corresponding disadvantage for any individual that would not pursue alliance with high-ranking individuals because of their sex. In matriarchal species such as elephants, bonobos, and orcas, females will aggressively pursue alliances with higher-ranking females and those alliances include sexual gratification.[46] In these species, a female who will not have sex with other females will quickly drop to the bottom of the social ladder and have no chance at successful reproduction.

Another idea regarding the advantage of homosexuality is found in the world of kin selection. Kin selection means that a trait can be favored through the enhanced reproductive success of relatives, who presumably also have the underlying genes. Genes for family cohesion and cooperation could be preferentially passed on, even when only some of the family members do the reproducing. In wolf packs, for example, some members forgo their own reproduction and instead help raise their siblings' offspring.[47]

This would apply quite nicely if it were found that homosexuals tend to help their families and contribute to their reproductive success.

It turns out that there is a study that has found this to be true. This work was done in one of the strangest mammals of all—humans—located in Samoa (the independent Polynesian nation, not the American territory). Canadian researchers working there found that gay men invest more time and financial resources into their nieces and nephews than straight men do.[48] Thus, kin selection could help explain the prevalence of homosexuality in humans.

You are probably thinking that this is a bit of a stretch to think that "nice gay uncles" offer so much of a survival advantage to their nieces and nephews that it would overcome any lost fecundity from their refusal to mate with women. I hear you. But remember two things. First, you are probably thinking of "nice gay uncles" in the modern world, lavishing gifts and picking up the kids from soccer practice. While humans were evolving, life was a constant struggle, and most children did not survive to reproductive age. Having another adult around to help with protection, food gathering, shelter fashioning, and predator lookout would have been really handy. It could have made the difference between life and death on more than one occasion, and voilà—there is the enhanced reproductive fitness.

Besides the data from Samoa, there is plenty of anecdotal support for the helpful gay uncle hypotheses. Many gay men opt not to have children themselves and instead spoil their nieces and nephews with both time and gifts. In the prehistoric era, this might have made a real difference in their survival rate.

Another hypothesis has emerged from studying homosexuality in humans. Several studies have found that female relatives of gay men have slightly more children than female relatives of straight men.[49] It is fair to say that this connection is still considered tentative, but it has been observed in a variety of ethnic groups in studies taking place in Italy, England, and Independent Samoa. So far, this increased reproductive rate has only been seen in the relatives of gay men, not lesbians. Interestingly, the increased fertility observed in the female relatives of gay men are those that are related to gay men via matrilineage.

Matrilines are lines of ancestry that follow only females and their off-spring. In other words, your mother's sisters are your matrilinear aunts. Your father's sisters are not. Your cousins that are the children of your mom's sisters are your matrilinear cousins, but the children of your mother's brothers are not, nor are the children of your father's siblings. Basically, matrilines are genetic relationships between two individuals in which the connection between them consists only of females. What the studies mentioned here have found is that the mothers, maternal aunts, maternal grandparents, and matrilineal cousins of gay men tend to have more children than those same relatives of straight men.

What does this mean? Two related hypotheses have emerged. One says that the genetic elements that are at play are expressed as homosexuality in males and hyper-heterosexuality in females. This claim seems to suggest that there are specific genes that encode sexual attraction to men, making men gay and women very straight. If you think this sounds odd, I agree with you.

A more accepted hypothesis is that, somehow, genetic elements responsible for gayness in men can lead to increased fertility in women. This makes a lot more sense to me. First of all, there is good reason to believe that sexual orientation in humans is only influenced by genetics, not strictly coded for. Other factors, including environment and even social factors in early childhood, likely play a role. Gestational hormones in utero have long been suspected, and some animal studies do support the notion that the sex steroids (testosterone, estrogen, progesterone) in utero play a role in development of sexual orientation later.[50] If this is true, a connection to fertility is easier to imagine since those same hormones can affect a woman's fertility.

There are two genetic factors that associate with matrilineal inheritance: the X chromosome and mitochondrial DNA. Since the very few mitochondrial genes seem to function only within the mitochondria itself, most believe the X chromosome is the prime place for genes that connect sexuality and fertility. Although a specific genetic element called Xq28 (named for its location on the chromosome) has been linked to male homosexuality,[51] subsequent studies failed to confirm this and the jury is still out on the matter of homosexuality and the X chromosome.

Regardless of the mechanism, would a connection between gay men and fertile women make sense in terms of evolution? It could, especially if you think about those gay uncles helping to raise the extra kids that their mothers, sisters, aunts, and cousins have. Although that seems a little too neat to me, it does make some sense and merits further study.

Another possibility is that some genetic element emerged that enhances female fertility. Then, purely by accident, it also increases the likelihood of homosexuality in men. If we assume that homosexuality causes at least some reduced fertility, this is a case in which the genetic element is being pulled in two directions, something that is not altogether uncommon in nature. Adaptations are often trade-offs, and maybe this is an example of that. In the case of male homosexuality, the genetic nature of it seems to be nothing more than a predisposition, an increased likelihood. Homosexuality does run in families, but only weakly so. Female homosexuality seems even less easy to pin down genetically, but that could simply be because less research has been done on gay women.

On a personal note, I routinely bounce back and forth between caring how and why people are gay. On the one hand, who really cares? Homosexuality is clearly a natural variation that is common among animals, especially mammals. It does not hurt anyone (although sometimes gay people are hurt by others because of it), and the vast majority of gay people say that they would not change their sexual orientation even if they could.

The scientist in me, however, is not satisfied with that. Of course we care about how people become gay, just like we care about how people get green eyes. Curiosity is in our nature; it is an essential part of who we are. I am not talking about "we scientists," but we humans. We are a curious species, always compelled to ask why things are the way they are. And that is a good thing—our endless curiosity compelled us to the moon, to the South Pole, and to the Higgs boson. Along the way, we invented radio, refrigerators, and the Internet. You know why? Because science.

The take-home point here is that sexual activity between members of the same sex is rampant throughout nature and quite pronounced in our own class, mammals, and our own order, primates. Thus, it is not surprising that it is so common throughout all populations and cultures of our species.

Whatever its reproductive costs, its benefits must outweigh them, or else it would not have persisted and flourished so far and wide. These benefits are likely to be similar in humans as in other animals because—you guessed it—we are not so different.

FREUD, OEDIPUS, AND INBREEDING

Sigmund Freud claimed that all little boys subconsciously want to kill their fathers and marry their mothers and termed this the Oedipus complex, after the classic Greek tragedy of Oedipus, whose main character unknowingly does just that. This flies in the face of the universal human taboo against incest. In all cultures across the globe, marrying and sexual activity among siblings or between sons or daughters and their parents is socially forbidden, and a vanishingly small percentage of people report any inkling of sexual interest in such endeavors. So, was Freud wrong?

Sex between close familiar relations (siblings, parents/children) is very rare indeed. The taboo is relaxed, however, when the relation is more distant. Sex and marrying among first cousins is not nearly as uncommon or as frowned upon as it is among siblings. Sex among second cousins is even less so. Beyond that, no taboo exists. In fact, the current Queen of England, Elizabeth II, married her third cousin, Philip Mountbatten. They are both great-great-grandchildren of Queen Victoria. Before you Yankees are tempted to mock the monarchists across the pond, keep in mind that Eleanor Roosevelt did not have to change her last name when she married Franklin D. Roosevelt, the thirty-fourth president of the United States. FDR and his wife were both Roosevelts, fifth cousins once removed. Also, FDR's parents were sixth cousins as well. While incest between siblings is unheard of, there is a strong tradition of "light inbreeding" throughout the aristocracy of Europe and the New World.

While it may not be widespread throughout society, marriage among cousins is at least acceptable in Western society and even more so among other world cultures. This "light inbreeding" does not seem to lead to any higher prevalence of genetic disease and was more common in preindustrial society than it is today. It was even considered preferable among the upper classes in order to keep the family wealth concentrated among a few noble lineages. Did purely social and economic forces drive that custom, or was there some biology behind it as well? Surprisingly, there are some data to indicate that humans may actually have an attractive force toward marrying their relatives. I am not talking about siblings, parents, or children, of course, but cousins.

First of all, it has long been accepted that, generally, humans tend to be most sexually attracted to members of their same race. This has been shown not just with marriage and dating choices, which could be explained by forces other than sexual attraction, but also with choices of prostitutes, erotic dancers, and pornography. I want to move very quickly past this point because it has been, and continues to be, tinged with racist indoctrination. However, I mention this because I find it interesting that the trend toward more integration and heterogeneity in the racial composition of North America and Europe has coincided with a reduction in the favored sexual attraction towards one's own race. This can be explained any number of ways, but one possible explanation is that the increasing exposure to people of other races during childhood has allowed more diverse sexual attraction to develop. Of course, this possibility will be inextricably intertwined with other social forces, but keep it in mind over the next couple pages.

Social scientists have also found that, at least for heterosexual couples, sexual attraction seems to favor those with similar eye and hair color.[52] In these studies, generally only Caucasians are examined since white folks tend to have a greater diversity of hair and eye color than other races. Of course, this correlation is not perfect, but the studies have found that people tend to be attracted to those who have the same or similar eye and hair color that they themselves have, at least more often than would be expected by chance.

This tendency for people to be more attracted to people with their same coloring has been discovered and documented enough times by different research groups that there is little disagreement in the scientific community that it is real. The question is, why would this be? Why would opposite-sex attraction tend to favor one's own physical features?

It turns out that that question frames the issue incorrectly. Heterosexual men and women are not most attracted specifically to their own hair and eye color, but that of their parents and other family members. Because eye and hair color are genetic and thus run in families, the fact that someone may be most attracted to their own hair color is incidental; it is really the hair color of those that raised them that most correlates to their sexual attraction. This subtle distinction has been shown most clearly by studying people who were adopted at birth, who are thus more likely than those raised by their biological parents to have different eye and hair color. (And race, of course.) One study in 2003 carefully explored this phenomenon and found that the sexual attraction of white heterosexual men and women most closely correlates with the eye and hair color of the opposite-sex parent, rather than those of self or the same-sex parent.[53]

This brings the whole issue into focus. Heterosexual attraction is at least in part the result of environmental imprinting during childhood such that adults become attracted to the features they saw in the opposite-sex individuals with whom they were raised. Fascinatingly, this is true not only with conspicuous features like eye and hair color, but also subtle things like ear shape and size, length of fingers, and distance between the eyes.[54] This makes perfect sense to me because sexual attraction can be very subtle. We cannot always pin down why we are more attracted to this person over that one. We might just say something like, "I don't know, she just has a better face." What we could be noticing, subconsciously of course, is that she has a general face shape most similar to the face shape of our mother and sisters. You might find that conclusion uncomfortable, but as my doctoral advisor often told me, "the data's the data."

The phenomenon of one's sexual preferences matching the physical features of close opposite-sex relatives has been reported many times. So far, nothing like this has been noted among gays and lesbians, which itself is calling out to be studied. What does this all mean? For me, it reveals a biological basis for the preference toward "light inbreeding" that we see in many human societies. Who looks most like our fathers and brothers (without actually being our fathers and brothers)? Our cousins!

This idea of Oedipal imprinting recently got a huge lift from an unlikely field of research—the study of the human genome. Researchers at the University of Colorado at Boulder recently examined genome-scale differences between individuals and their spouses. To keep things simple, they established the most homogeneous population possible by sticking to a single race, Caucasians of non-Hispanic ancestry (big surprise), and they included only heterosexuals. In 825 such couples, they examined 1.7 million single-nucleotide polymorphisms (SNPs). An SNP is the tiniest possible genetic difference between two people. What the researchers found was that married couples tended to have fewer genetic differences between them than you would expect at random. In other words, if you selected a guy and a girl at random from this group of 1,650 people, they would, on average, have more genetic differences than the married couples have to each other.

This study did not probe exactly what the genetic similarities or differences were, nor did they ask how we humans can possibly know who out there is genetically similar to us and who is not. Those are both very difficult questions. Nevertheless, the results of the study are rather convincing, given the very large and homogeneous population that was studied. There seems to be something real going on here, some force that pulls together people with similar DNA. We have always known that human mating is nonrandom. People tend to marry people of their same socioeconomic status, educational level, race, ancestry, political persuasion, and geographic location. However, this new study found that, even within those specific groups, there is another level of attraction at work—we are drawn to people with similar genetic information.

It turns out that Freud was at least partially right. Subconsciously, many of us are somewhat attracted to our mothers or fathers—but not to them personally. We are attracted to the features that they have when we see them in other people. Because we will likely be attracted when we meet someone that looks or acts like our close relatives, attraction among cousins is perfectly understandable and, in fact, expected.

This creates a conundrum because this would seem to argue against a taboo against incest. Are we conditioned to avoid incest purely by social forces, or is there something biological at work? How is it that we can be attracted to the features of our parents and siblings, but not our parents and siblings themselves? Most of us are not attracted to our parents or siblings.

Where does the incest taboo come from? The answer again seems to be social imprinting during childhood. Edvard Westermarck, the so-called father of sociobiology, first initiated the study of incest avoidance in humans and animals. He began his work by studying the marriage practices of humans in many cultures around the world and found that, in addition to the already well-known close incest taboo, people tend not to marry individuals with whom they closely associated as children.[55] This phenomenon became known as the Westermarck effect and is almost universal throughout human cultures and time periods. It is not just our siblings that we avoid: we humans tend not to marry anyone with whom we grew up closely. (Exceptions to the incest taboo are found among the ruling class of several ancient civilizations. This may have extended to commoners in ancient Egypt.)

Of course, spousal choice is complicated by the fact that, in many parts of the world, spouses are actually chosen by people other than the betrothed themselves. This was especially true when Westermarck was doing his work in the late nineteenth century. Further still, perceived sexual attractiveness does not always feature prominently into the decision-making process for arranged marriages, at least not ostensibly. However, given that potential brides and grooms do have some voice or veto power, there is plenty of room for the preference for avoidance of close childhood peers to express itself among the would-be spouses.

Not long after he first published his ideas, support for the Westermarck theory began to roll in from a wide variety of cultures and contexts around the globe. The strongest such support came from a now-classic study performed among residents of an Israeli kibbutz.[56] The kibbutz, or collective farm, is an ideal context for studying this phenomenon because, within the large settlement, children are often reared in small communes. Within the communes, several families live together and behave as one large family even though they are not usually related to one another. The children in the commune are mostly raised together in a collective effort.

In this study, researchers examined both marriage choice and premarital sexual behaviors and found that individuals specifically had sex and/or contracted marriage with peers from outside their specific commune—in other words, those they were not raised with. The statistical analysis was convincing. Out of three thousand marriages that took place among second-generation kibbutz residents, only fourteen involved two individuals that had grown up in the same commune.[57] The avoidance of childhood peers was much stronger than random chance could explain, even though there was no cultural taboo against marrying someone from the same commune. An earlier study had reported an even lower percentage of intracommune marriages, but was not as rigorous or quantitative in its approach.[58] Something about the experience of growing up with someone, even a nonrelative, made desire for sex or marriage unlikely.

There have been challenges to the Westermarck-based interpretation of the data from Israeli kibbutzim.[59] An alternative hypothesis is that low rates of intracommune marriages may be the result of tendencies toward the maintenance of group cohesion. Many of us have personally experienced the disruptive consequences when two people in a group of friends begin (and end) dating. The avoidance of dating and thus marriage among close childhood friends could be reflective of that, and would be a sociological phenomenon, rather than a biological one. Importantly, while the opponents of Westermarck quarrel with the mechanism of the incest taboo, they do not doubt its near universal existence in human society.

As I mentioned in chapter 1, playing and growing up with someone creates a special bond that long endures. Part of that bonding experience, it seems, is the imprinting against mate selection. As children, the people that we spend the most amount of time with are family. This is as true now as it was in our hunter-gatherer days. Thus, in order to avoid heavy inbreeding, our biology may have programmed us to be sexually averse to those that we spend a great deal of time with during childhood, when social imprinting occurs.

Humans are not alone in this. Westermarck quickly turned to other animal species and found a similar pattern.[60] Many species of animals tend to avoid reproductive sex with their siblings and others with whom they were closely reared. In almost all primate species that live in packs, families, and harems, males leave the group when they begin their sexual lives. What happens after that is different among different species. In gorillas, for example, the young males will roam on their own for a while in search of a harem to take over. Most fail and die in the process, which is why harems have one male and several females.

In chimpanzees, most often it is the females that join another troop when they are old enough to roam.[61] Still, those that do stay or those that are raised in captivity will not reproduce with males that they grew up with, siblings or not. In the wild, some females go even further and take short trips during their estrus period to have sex with males from other troops. Interestingly, chimps that were raised together, even siblings, will indeed have sex with each other but not reproductive sex. Both males and females know when the fertile reproductive period is and they almost always choose to out-breed at that time.

It turns out that, in her wisdom, nature does not rely on social imprinting alone to help guide us away from too-close relatives when choosing sexual partners. There are even more subtle forces at work. In 1974, Lewis Thomas first suggested that animals might be able to detect genetic features of potential mates using their sense of smell.[62] He was later proven correct when it was discovered that mice can detect genetic differences in the MHC proteins using smell and will selectively breed with mice that have different forms of MHC than their own.[63] Presumably, this is a means to promote genetic diversity and avoid inbreeding.

The MHC proteins play an important role in the immune system of vertebrates by helping us "mark" our own cells and proteins and protect them from self-attack. MHC proteins are present on almost every cell of our bodies and have important functions in our fight against viruses, bacteria, and other pathogens. However, they also mediate the rejection of foreign cells, hence their name: MHC stands for major histocompatibility complex. When organ and tissue donors are sought for transplants, doctors search for a "match," which means someone with the same or very similar MHC proteins. If a transplanted organ has a different set of MHC proteins, it will be rejected by the recipient.

In our modern era of organ transplantation, we must perform an elaborate laboratory test to determine if two people have matching MHC proteins. However, nature figured out how to do this eons ago. Animals can actually detect the MHC proteins of other animals by their smell. While this was puzzling at first, with the discovery that MHC proteins play important roles in olfactory perception per se, a possible mechanism is now clear.[64]

The most fascinating part of this is that animals will use the olfactory detection of MHC proteins to help them select mates that are genetically different from them. This phenomenon has now been observed in hundreds of species including mice, several kinds of fish, lizards, birds, and in lemurs, our primate cousins, and is lost when the sense of smell is impaired.[65] This phenomenon is widely found but has varying strength. Some species have extremely strong preference for mates with different MHC proteins, while others do not really seem to care.

What about humans? You guessed it. We are no different. The first experiment of this type was done by asking forty-four men to take home a clean T-shirt and wear it for two days with no deodorant, antiperspirant, perfumes, or cologne. Then, they returned the T-shirt, each nicely saturated with the natural body odor of the man who wore it. The T-shirts were put in a box with a hole cut out for the aroma to escape. Forty-nine women were then asked to smell the odor coming out of some of the shirts (and an unworn shirt as a control) and select the ones that they found most and least attractive. There were no clear winners or losers; each male scent had its fans and its objectors. However, with striking consistency, the females favored the scent of men whose MHC proteins were most dissimilar to their own.[66] Some of these women even admitted that the smells they liked, those from the men with dissimilar MHC proteins, reminded them of current or previous boyfriends/husbands.

There is another wrinkle in this MHC-odor preference phenomenon. The MHC preference of mice and some other species switches when they are pregnant. Instead of preferring to be with mice of different MHC proteins, they tend to associate with mice of similar MHC profiles.[67] Similarly, in the study with the human women smelling the sweaty T-shirts, those on birth control (which somewhat mimics pregnancy in having high levels of estrogen and/or progesterone) behave in exactly the opposite way as well: they prefer the odors from men with MHC proteins that are similar to their own.[68] What to make of this?

The simple interpretation is that female mammals are sniffing out potential mates when they are not pregnant, but preferring relatives when they are. But why? When one is pregnant, there is little need to look for potential mates. Instead, the preference for close relatives while pregnant is likely a desire for safety. Male relatives (father, brothers, sons) are much less likely to kill you or your baby than are strangers. As noted in chapters 7 and 8, there are reasons to be cautious around nonrelatives. There is value in seeking genetic diversity when you are looking to procreate, but once that is done, it is best to retreat to the safety of family. This is as true in mice as it is in humans.

Now it is time to clean up the apparent contradiction that has been brewing. First, I explained how humans are attracted to physical features that match those of their close opposite-sex relatives and how married couples tend to have overall greater genetic similarity to each other than to otherwise similar people. Then, I explained how humans and other animals are attracted to genetic diversity when it comes to the MHC proteins. This seems like I am saying two exactly opposite things and that they cannot both be true.

Not exactly. The preference that many people have for the physical features found in our family members is a form of biological nationalism (or, more darkly, racism). This speaks to the subtle ways that "the selfish gene" tends to promote its own success. In social animals, this leads to the creation of packs and herds and the creation of an "us versus them" mentality that we will discuss in more detail soon enough. Although these forces help shape highly competitive species due to their natural tendency to promote elitism and exclusivity, there is a downside: elitism and exclusivity can lead to heavy inbreeding. Heavy inbreeding leads to a great poverty of genetic diversity. Low genetic diversity makes an individual vulnerable to all kinds of threats, most especially pathogens such as bacteria and viruses. Without some diversity as insurance, one good virus can wipe out an entire species in no time.

This is where the MHC proteins come in. By preferring to mate with someone that has a different set of MHC proteins, we mammals have a built-in mechanism to resist the rest of our urges to become homogenous. Part of our sexual attraction is fueled by the preference for light inbreeding, which is good for the propagation of the family genes, and another part is fueled by the preference for genetic diversity in our MHC proteins, which is good for the survival of our offspring themselves. Remember that MHC proteins function in our immune system. It is not surprising that vertebrates have evolved special means to ensure that our immunological toolbox is stocked with the most diverse possible tools. In fact, in a very real sense, the quality of an immune system can be measured by the diversity of threats that it can handle. An individual with a monolithic array of MHC proteins is surely doomed.

There is something of a contradiction in these two natures of our sexual attraction, one toward those like us, and one toward those different from us. However, this is just the kind of push-pull relationship that we see throughout nature. We are always being pulled in multiple directions and natural selection usually settles on some kind of compromise. In birds, larger wings make better flyers. So why is it that all bird species do not have enormous wings? Because smaller wings are easier to tuck away while on the ground and do not require as much metabolic energy to build and maintain. Two forces pulling in opposite directions.

Sexual attraction is another such example. Inbreeding is great for rapid selection of an elite and well-adapted form, but out-breeding is necessary to maintain genetic diversity, without which we would quickly succumb to genetic diseases or infections. Freud and Westermarck were both right.

FURTHER READING

Gray, P. B., and J. R. Garcia. "Evolution and Human Sexual Behavior." *History and Anthropology* 24 (2013): 513–515.

Wright, Robert. *The Moral Animal: Why We Are the Way We Are; The New Science of Evolutionary Psychology.* New York: Vintage, 1994.

6

THE AGONY OF GRIEF

RIEF, ONE OF THE MOST unpleasant emotional states, is the response to loss. This is not to be confused with depression or plain old sadness. Grief specifically refers to the emotional pain suffered upon experiencing a profound loss. It is most associated with the loss of a loved one, but it could also be the loss of a job, home, health or bodily ability, valued possession, or a relationship. For simplicity, in this chapter I will focus almost exclusively on the grief caused by the loss of a loved one. That is usually the deepest grief that we experience and is the most easily observed in animals.

Grief is a highly personal emotion, experienced very differently among individuals. No two people experience grief the same way or with the same intensity. Some people are able to dust them themselves off and get on with their lives relatively quickly, while others wallow in misery for months or even years. Psychologists have not fully figured out the factors that lead to this high degree of variability. Regardless, everyone experiences an intense period of sadness and despair upon the loss of a loved one.

Before we consider the existence of grief in animals, we should admit up front that, at first glance, grief does not serve to perform any obvious biological function or provide any readily identifiable benefit to the individual or the human race. How is it good for humanity that we feel so lousy when someone dies? Perhaps it would be better if we were built to simply move on after a death in our family. It does no one any good to be despondent for an extended period of time. This is going to be a tough one to explain.

As with everything else, I think the best way to discover how and why humans experience grief is to explore how and why animals experience it.

DO ANIMALS GRIEVE?

One of the most famous examples of animal grief is the story of a wild chimpanzee named Flint, observed and documented by pioneering primatologist Jane Goodall. Flint was the third child of Flo, the highest-ranking female of a chimpanzee troop in Tanzania. As Professor Goodall describes it, Flint had all the trappings of a spoiled youngest child (something I am a bit familiar with myself). He clung to his mother incessantly and refused to do things for himself. He threw tantrums and acted much younger and more helpless than his true age of four years old. He preferred the constant company of his mother and played very little with his peers.

Flint even had trouble weaning, breastfeeding far longer than normal. Flo was in her forties when Flint was born and perhaps she just did not have the energy to struggle and acquiesced instead. Whatever the reason, Flint was a mama's boy if there ever was one.

When Flo had yet another child, Flame, Flint was displaced as the baby in the family and suffered what can only be described as emotional trauma. He regularly threw tantrums in an attempt to nurse, especially when Flo held Flame in her arms. Flo was not able to support both children and pushed Flint aside. He became socially withdrawn and depressed, and he lost considerable weight. In other words, he appeared to be in a state of grief.

However, his poor fortune was short-lived when Flame suddenly disappeared. Professor Goodall never fully determined Flame's whereabouts, but it seems that she somehow died at the tender young age of just six months. No longer in competition for his mother's attention, Flint regained his energy and enthusiasm. He eagerly went back to his mother and began nursing and eating again. He regained his health and vitality.

Interestingly, Flo took her spoiled child back into her close care with less resistance than she had shown in the time leading up to Flame's death. Perhaps she was comforted in her loss by caring for Flint in the infantile fashion that he preferred, which was more age appropriate for the lost child than for Flint himself. Flo gave up trying to force him to be independent and let her five-year-old child act (and breastfeed) like an infant.

Three years later, Flo died at an estimated age of well over fifty. Flint was devastated. He completely withdrew socially and stopped eating. Other chimps tried to coax him back into the fold, but he refused. He would often sit, awake but unmoving, for hours. At the age of eight-and-one-half years, Flint died. Physiologically speaking, he likely died of dehydration or undernourishment. Here is an excerpt from Professor Goodall's book, *Through a Window*:[1]

> Three days after Flo's death, Flint climbed slowly into a tall tree near the stream. He walked along one of the branches, then stopped and stood motionless, staring down at an empty nest . . . one which he and Flo had shared a short while before Flo died. . . . Then he suddenly . . . raced back to the place where Flo had died and there sank into ever deeper depression. . . . Flint became increasingly lethargic, refused most food and . . . fell sick. The last time I saw him alive, he was hollow-eyed, gaunt and utterly depressed, huddled in the vegetation close to where Flo had died. . . . The last short journey he made, pausing to rest every few feet, was to the very place where Flo's body had lain. There he stayed for several hours. . . . He struggled on a little further, then curled up and never moved again.

You may want to have some tissues handy while reading this chapter.

Other examples of animals grieving come from owners of multiple pets. As an example of countless such anecdotes from pet owners, I offer a story told by Barbara King, author of *How Animals Grieve*, about two Siamese cat sisters, Willa and Carson.[2] These two sisters lived together with their owners for fourteen years with no other pets in the house. They were inseparable, sleeping and sunbathing together, sticking tightly together when uneasy about unfamiliar guests visiting the home, and taking their meals together.

When Carson died after a short illness, Willa reacted immediately. She showed the typical agitation and searching behaviors that she had exhibited when the pair had experienced temporary separation for vet or grooming visits. This time, however, Willa was different. She searched and researched the entire house, especially the places where she and Carson had passed so much time together. She showed no interest in food or play, and, perhaps most disturbing of all, she let out incessant and haunting wails that her owners had never heard her make before. She was visibly agitated and found no comfort from physical affection from her owners, and the behavior went on for several weeks.

To anyone who has experienced the death of a pet with other pets remaining, that story will probably ring familiar. When we have multiple pets, they form a pack of sorts, even sometimes a mixed-species one. The loss or temporary absence of one member of the pack is felt keenly by the other members. Even cats and dogs that do not normally get along with each other will often react to each other's deaths with sadness and anxiety. This is undoubtedly made worse if and when some pets sense and "catch" the sadness of their owners (see chapter 3 on empathy), which probably helps to signal the finality of death, compared with a temporary absence for whatever reason.

Dolphins are known as one of the most intelligent, most socially complex, and, yes, one of the most emotional of animals. It has even been shown that they recognize themselves in mirrors, a phenomenon previously thought to be unique to humans and some apes.[3] Not surprisingly, they have been known to display various behaviors that appear to be grief.[4] Dolphins will carry lost loved ones, particularly juveniles, on their backs, sometimes for days. During these days of grief, they will not eat or sleep, remaining steadfastly dedicated to the dead dolphin they are carrying. Various species of ocean dolphins have been spotted doing this. Because these instances are observed in the wild, scientists have not confirmed what we all suspect: that these heartbreaking scenes are mothers carrying their dead children. There are many heartbreaking images and videos that can be on the internet by searching for "dolphin carries dead calf."

In one such video from China, onlookers watch as an adult dolphin swims and battles the waves while carrying the dead juvenile on its snout or back. The dolphin drops the carcass no less than five times and swims to retrieve it. It is unclear where he or she is headed with the body, but she struggles in the surf for an extended period to arrive there. The adult dolphin surfaces often and swims for extended lengths of time with her head and the lifeless body above the surface of the water. Wild dolphins do not frequently swim with their heads above water; they usually surface by swimming in an arc-shaped motion so only their dorsal fins and blowholes are exposed for an instant. Occasionally they jump, but this particular dolphin appears to be lifting the lifeless body almost completely out of the water for as long as possible before it slides off. Could this have been a desperate attempt to revive the dead child by bringing it to the surface to breathe?

It is not entirely clear that this dolphin was grieving, but this incident is not unique. Scientists and others have documented many such stories of dolphins carrying carcasses, sometimes for days at a time.[5] Especially because the bodies are always those of juveniles, it seems plausible that a mother dolphin is experiencing the trauma of losing a child that she recently bonded with. Remember our friend oxytocin? Never is it more active than in a nursing mother, which helps create the vicelike grip of attachment in mammals. It seems that these poor dolphin mothers refuse to let go.

Wolf packs have notoriously strong bonds among members. The social structure is so organized that some members of a pack will forgo sex and reproduction and instead contribute to the survival of relatives' children. This degree of cooperation requires very strong social bonding, which is probably why wolves were chosen as the ultimate companion animal for early humans. As wolves evolved into dogs, the instincts for social bonding were exploited so that the dogs came to see their human families as their packs. However, it is important to keep in mind that the impressive social bonding of dogs preexisted in their wolf ancestors.

So strong are the bonds among wolves that the animals actually miss each other when they are separated. This has been recorded anecdotally for some time, but recently, a group of scientists in Austria found a way to explore and document it experimentally.[6] In this study, wolves from a social pack living in a wild animal preserve were subjected to temporary separations by the removal of individual wolves from the pack for short periods. During the separation, the scientists observed and documented how the remaining wolves reacted—their howling, in particular.

First of all, it was clear that the overall degree of howling increased in the pack when one member was missing. No surprise there; this is just another example of the social cohesion of the wolf pack. The really interesting thing that scientists found was that the wolves that did most of the howling were those that were the most closely bonded to the missing member. In a wolf pack, just like a human social group, not all relationships are the same. A given wolf will be more closely bonded to some wolves than she is to others. Some wolves are outright rivals, and dominance challenges are frequent. Other pairs of wolves are closely bonded to each other as siblings, parent-child, mates, or even just friends. This can be measured by simply documenting how much time individual wolves spend with each of the other wolves in all of the pair-wise combinations. This phenomenon can also be shown by documenting aggression events, which then reveals instances of reduced bonding among a given pair.

The study found that when a wolf is missing, the wolves that are closely bonded to that wolf will howl at regular intervals. As the separation endures, the howls become more frequent and the wolves become more distressed. We could describe this behavior in various ways that deemphasize the emotional nature of the bond between the missing wolf and her friend/mate/sibling, but must we? It seems more reasonable to conclude that the wolves are concerned for their lost companion. The wolves suffer the withdrawal of a social bond and react by trying to reestablish it. While this is not grief, per se, it captures one of the essential components of grief: loss. This study confirms that wolves are keenly aware of the loss of a member of their group, and, even more important, they feel it much more deeply when it is someone they are close to.

The likely explanation for this behavior is that the wolf brain is built to sense the sudden loss of a social bond and respond by calling out to the lost member. This behavior conveys obvious biological benefits by promoting group cohesion: the howler gains by reconnecting with a social ally, and the separate wolf gains by finding its way back to the group. Many dog breeds are known to suffer from separation anxiety when left alone, which is probably due to the same brain event—a response to deprivation of a social bond.

LESSONS FROM ELEPHANTS

Elephants are usually gentle animals, mainstays in zoos and circuses, and beloved of wildlife activists due to their intelligence, adaptability, apparent emotional complexity, and exceptional memories. Biologists have long appreciated the incredibly complex social interactions of elephant herds. Herd mates bond with each other in a tightly knit community and, given the long life span of elephants, develop decades-long attachments. This combination of traits seems to be the perfect cocktail for the experience of grief. Indeed, elephants are the animals that serve as perhaps the most striking examples of animal grief. I will share just a few stories to illustrate.

Rennie Bere, a game warden in Africa, tells many stories of elephants grieving for their dead, including one in which a grieving mother carried her dead child with her migrating clan for days.[7] She had to set the child down periodically to eat and drink; she sometimes needed to adjust her stance and method of carrying. All of this slowed her down considerably. What did the rest of the clan do while this burdensome member trailed behind while she was grieving or was possibly even in denial? They slowed down and accommodated her.

In her book, *Coming of Age with Elephants*, Joyce Poole tells several stories of elephant grief.[8] She describes a mother mourning over a stillborn child. She remained with the body for two whole days until lionesses grabbed the carcass and took it away for their meal. During her vigil, she repeatedly nudged her dead child with her trunk and made soft and low sounds that can only be described as sobs. Tears ran down her face.

This "crying" behavior has been documented in elephants many times before and since. Many scientists and zoo and nature preserve workers believe that elephants cry. In response to tragedy, they emit soft moans, and their tear glands begin to produce so excessively that the tear ducts cannot absorb it all and the tears spill out onto their faces. That sure sounds like crying to me.

In elephants, it is not only parents of dead children that weep. In early 2013, a baby elephant was born in a nature preserve in China. For some reason, the child was rejected by its mother, who attacked him not once, but twice. Animals can be disturbed psychologically, just like humans. Unfortunately, rare cases of child abandonment and rejection are something that humans have in common with other animals. In the case of this unfortunate young elephant, however, employees of the nature preserve had the rare and terrible opportunity to witness the elephant child's reaction to the abandonment. It was not pretty. The young elephant cried inconsolably for hours and hours. He was not hungry; he did not need shelter or warmth. He needed the love of his mother, and he did not get it. In response, he wept. The article has pictures and I encourage you to find it online. You can see the tears streaming down his face. You can almost hear his whimpers.[9]

A young pygmy elephant named Borneo grieved the death of his mother in a preserve in Malaysia. The young calf continually nudged his mother's body and let out cries of frustration and stress that were much different than the typical cries for attention we hear from young elephants. Workers in the preserve offered water and milk, but little could distract the young calf. Eventually, they had to forcibly take the calf away so that they could remove the corpse of his mother.

The strong bond between elephant parents and children can even form in situations of adoption. In 1999, a young pregnant elephant, later named Champakali, was moved from a national park in south India to a nearby zoo in a city called Lucknow to protect her from poachers and other dangers of life in the wild. At the zoo, a middle-aged female elephant named

Damini almost immediately bonded with the new addition. She cared for Champakali and doted on her constantly. The handlers had never seen anything quite like it. Damini had previously been something of a loner with only weaker bonding to the other elephants in her group. This all changed when Champakali entered the picture and the two became inseparable.

Tragically, despite the care of veterinarians, Champakali died in childbirth, along with her calf. The loss rocked Damini to the core. She immediately stopped eating. She cried visibly and audibly and, over a period of three weeks, reduced her consumption of food and even water. She collapsed and lay on the ground for a week as veterinarians attempted to care for her. She died four weeks after her surrogate daughter and was buried alongside her.[10]

An elephant named Patience trampled and killed one of her handlers in a Missouri zoo during an outburst of frustration.[11] She had been agitated and acting out for three days. What was causing her to be so out of sorts? The matriarch of her group had recently been euthanized due to severe kidney disease. Because it seems pretty unlikely that elephants understand the concept of merciful euthanasia, murder would be the more likely interpretation by the elephants. It is not clear from news reports if the handlers had "waked" the dead matriarch to allow the rest of the group to understand, grieve, and accept her death, so it is possible that Patience's act of aggression was less out of grief and simply an attempt to discover what they had done with her beloved leader. Either way, the death of her companion left her in great distress.

LESSONS FROM PRIMATES

In a zoo in Münster, Germany, there is an adult female gorilla named Gana. At the age of eleven, she gave birth to her second child and first son, Claudio, and cared for him dutifully. Suddenly, the caretakers noticed that Claudio was losing weight, not eating much, and showing little energy.

Before the veterinarians could examine him, he suddenly died. An autopsy later revealed the cause: a congenital heart defect. He was three months old.

Gana did not take this well. As described by many handlers and visitors to the zoo, she was "consumed with grief."[12] Gana frantically stroked the child, pick up his lifeless arms and legs, and stayed with him for several hours, caressing and prodding him to no effect. This, in and of itself, is not uncommon among gorillas and other apes, but what happened next was indeed surprising. Gana finally moved from the spot where her son had died, but she did not leave him behind. She placed him on her back and carried him piggyback, just as she had when he was alive. She went about her business, carrying him around the compound either on her back or in her arms, stopping frequently to probe and prod him, as if to see if he had returned to life.

The zookeepers were unable to retrieve the dead child from Gana for several days as she fiercely guarded her lost son. She went mostly without food during her days of mourning and did not interact with other members of the group. The only sounds heard from her were whimpers and grunts, unless a handler or another gorilla approached, when she would growl. Only after several days did she finally lay down her decomposing son and move on.

While this scene captivated Germany, it was nothing new to those who study gorillas in the wild. Gorilla parents have been known to mourn dead children and carry them around, sometimes for weeks. Gorilla pregnancy is even longer and more demanding than human pregnancy, and raising children is an exhausting, years-long endeavor. Further, gorillas only typically live for forty to fifty years and do not become sexually mature until they are eight or nine. Altogether, this means that gorilla mothers typically only give birth to singlet children, spaced many years apart. Thus, the mother-child bond is extremely strong in this species. Where there is great love, there is great loss, so these animals grieve hard. It is very difficult for these mothers to let go.

The phenomenon of a mother refusing to part with her dead child has been documented very extensively in baboons as well. Baboon mothers will carry their dead children around, sometimes for days. They continue to groom them, stroke them, attempt to nurse them, and shake them gently, as if to wake them. Only when the bodies begin to visibly decompose do the mothers finally leave them behind and move on. There are many heartbreaking videos of baboon mothers carrying their dead children on YouTube. Explore them if you are in the mood for a good cry.

Macaques have also been known to wallow in grief upon the death of a troop member. A worker in the Don Chao Poo Forest Research Center in Thailand captured on video the grief of a mother macaque cradling her dead child. In the video, which is posted on YouTube by the user phanamonkeyproject, the mother continues to offer food to her dead child, which she carries around like a limp doll. At first, she runs away as other monkeys approach, as if to protect her child or her food. Then, she allows one to approach. The visitor is not interested in her food. The other monkey tries to console the mother, stroking her gently, from behind, and caressing the dead baby. The mother seems to accept the comfort for a while, and then runs off again.

In another video available on YouTube, a clip from the television show *Wildest India*, a baby macaque is killed during a fight between two rival macaque troops. After the fight is over and the troop realizes that they suffered a casualty, the dead child's mother can be seen gently caressing and rubbing the body. She emits cries of distress and anguish and carries her dead daughter around for hours. The mood throughout the entire troop becomes mellow, somber even, and other members come to offer comfort periodically. After the mother finally lets go and leaves the baby behind, another juvenile of roughly the same age comes over for one final good-bye. He touches her gently, sits quietly next to her for a few seconds, and then lets out a scream before leaving the body behind.

Finally, I turn to our closest relatives, the chimpanzees. We have already discussed the case of Flint grieving himself to death over the loss of his mother. While it is tempting to chalk that case up as an outlier—an extreme case due to an exaggerated mother-son bond and incomplete adolescent development—there are other examples of chimpanzee grief even among nonrelatives.

At the Sanaga-Yong Rescue Center in Cameroon, there lived a chimpanzee named Dorothy.[13] Before reaching the rescue center, she had lived a hard life for decades at the hands of an amusement park. Despite that, she became an especially gentle and friendly resident of the rescue center for her last eight years, and she was popular with both the humans and her fellow chimps. She never had a child of her own, but "adopted" a young orphan male named Baboule. Her love and attention coaxed Baboule out of a deep sadness and shyness. As he gained confidence and grew strong, he took his place in the social order of the troop and will likely be a candidate for alpha male someday. Dorothy continued her practice of fiercely protecting Baboule until long after he had outgrown her. She also had many friends and would generously groom others while only rarely asking for reciprocal grooming.

Dorothy died of heart failure at an unknown but certainly advanced age. All of the chimps at the facility fell into a state of agitation and lethargy. Her adopted son Baboule, now a large adult, was hit the worst and wallowed around on the ground. There was very little eating in the troop for a few days. Even the alpha male, Jacky, screamed on the ground and had to be comforted by others.

A couple of days after Dorothy's death, the employees buried her in the animal graveyard near the chimpanzee enclosure. They invited residents of the nearby village to view her body and pay their respects, and many did so. The chimpanzees all lined up at the fence and stared at her body in almost complete silence. (Monica Szczupider snapped a remarkable photo of the scene, which you can find online by searching for "Dorothy's funeral" and Monica's name.) Because chimpanzees are not often quiet or still, the memories of that serene mood continue to haunt the workers at the rescue center to this day. After Dorothy's body had been wheeled past the other chimps, the staff proceeded to the graveyard for the burial. The chimps began to scream and carry on, so they brought Dorothy's body back to the fence for one last good-bye. After another day or two of mourning, the troop gradually began to return to their normal routines.

RECOGNIZING GRIEF

Some are hesitant to use the term "grief" when discussing animals. In response to the many stories above, they might offer one of the following explanations. First, they may say that the altered behavior of the animal is actually anxiety and uncertainty that is associated with a sudden change in its social routine. Animals need time to settle into a new routine and will be uneasy until they do. Second, they might say that "grieving" animals are really just suffering from behavioral symptoms caused by the rebalancing of their neurochemistry. Hormones in the brain will be affected by the sudden removal of a conspecific to which they had been attached.

These two things, so goes the reasoning, lead to stress, mediated by neurotransmitters and hormones that we discussed earlier (cortisol, norepinephrine, and so on). The symptoms of stress include reduced energy and appetite, restlessness and disrupted sleep patterns, and loss of interest in play (because, as discussed in chapter 2, play is only something an animal will do when not stressed). In essence, these critics would say that the animals described in this chapter are not grieving; they are simply stressed.

In my opinion, these criticisms are actually nice, concise descriptions of what grief is. I am quite happy with those behavioral descriptions of the animal response to great loss. In fact, I am even comfortable applying those statements to human behaviors that are associated with bereavement. The only thing missing is the account of the tremendous internal pain—the despair associated with grief—only because we cannot ask them to describe how they are feeling.

However, if we were able to know what they were feeling, why would we think that they would not experience emotional pain upon the death of someone with whom they have bonded? They certainly act as if they do. Let me put this another way. As you read about the grief behaviors of animals throughout this chapter, are you struck by how human they seem? If we were to describe the human behaviors—without talking of the emotion, just the behaviors—after a close loved one has died, it might be very similar to what we observe in animals.

Grieving humans often eat less and lose weight. They appear lethargic. They lose interest in things they normally enjoy: in other words, they refuse to play. They will withdraw socially from friends and family. They will visit places that remind them of their lost companion—their bedrooms, favorite places, and so on. They will visit the graveyards or places of death of the deceased. These are all reminiscent of what elephants, dolphins, chimps, and gorillas do. If there are stark similarities in our grieving behaviors, perhaps the default assumption should be that the underlying emotional experience is similar also?

The idea that animals experience grief was not controversial in previous eras. Ivan Pavlov stated in 1927: "Different conditions productive of extreme excitation, such as intense grief or bitter insults, often lead, when the natural reactions are inhibited by the necessary restraint, to profound and prolonged loss of balance in nervous and psychic activity."[14] Pavlov was talking about dogs.

In the late nineteenth and early twentieth centuries, the explosion of biological research seemed headed in the right direction of recognizing our close emotional kinship with animals, a movement launched largely by Darwin himself. However, around the middle of the twentieth century, a strong repulsion toward qualitative research and anthropomorphism sprung up in the scientific community, and the serious study of animal emotions was swept away with almost puritanical fervor.[15] The stripping away of common sense drastically crippled the study of animal behavior. Thwarted and ridiculed at every turn, biologists interested in animal emotions often simply gave up and shifted their focus to other research matters.

Luckily, the study of animal minds and emotions, and thus a more robust and accurate view of animal behavior, has seen a resurgence in the study of biological sciences over the past two decades. To all of those who rejected and continue to reject the view that animals really feel anything in their minds, I retort, yet again, with words from Charles Darwin himself: "Who can say what cows feel, when they surround and stare intently on a dying or dead companion[?]"[16]

Of course, a purely behavioral look at grief does not capture the full inner experience—the pain of grief—but that is all we can do when it comes to studying grief in animals. Further, no one I know of has ever said that the inner experience of grief is the same between humans and other animals. We can fully admit that grief, like all human emotions, is likely much more intense for humans because of our highly evolved and complex cognitive abilities. We have the gift of introspection, which is more like a curse during moments of grief. It seems likely that introspection is what makes our grief so terrible. When we lose someone we love, it would probably be a bit easier if we were less aware, less thinking, and less feeling. If we simply forgot about our loss after a few days, we would probably be able to move on with little trouble.

Do animals, especially ones with smaller, simpler brains, experience grief less profoundly than we do? Probably. In fact, their more rudimentary experience of the loss of attachment is probably closer to how grief first evolved, so of course it would be simpler.

I offer a parallel. The sense of vision first evolved as a small patch of pigmented cells in a small marine invertebrate. Pigments, molecules that absorb specific wavelengths of light, allowed these cells to sense light. These small invertebrates could then respond to that light, and so began the evolution of vision. Yes, earlier animals had a much simpler sense of vision, but they were well on their way. It is the same with grief. Other animals have smaller brains, but they do appear to experience grief. Furthermore, while we can admit that most animals probably do not experience grief as intensely as humans do, let us not forget Jane Goodall's report of Flint, the chimpanzee. When was the last time you heard of a human starving to death from grief?

THE ORIGIN AND PURPOSE OF GRIEF

Clearly, the pain of a lost attachment is detrimental to the health of the individual. The pages of this chapter are filled with stories of animals detaching socially and refusing food and sometimes water—even to the point of death. One study even found that grieving sperm whales develop measurably weaker teeth attachments to the point of actually losing teeth![17] In humans, the brain activity associated with grief can actually be observed in brain scans. Interestingly, some of the biochemical changes associated with grief in humans are those normally associated with inflammation, such as during infections or allergies. Scientists have even been able to detect and measure this grief-associated inflammation in the saliva of bereaved persons.[18] Clearly, grief is not good for you.

So if grief drastically lowers the survival chances of the bereaved, how did it ever evolve in the first place? Currently, the most widely accepted explanation of the biological origin of grief is that it is nothing more than an unavoidable side effect of attachment. As we discussed in chapters 4 and 5, social animals have evolved hormonal mechanisms to promote attachment. Attachment is being drawn to someone and yearning to be in his or her presence as much as possible. Attachment requires that you feel the absence of your loved one (and then try to reestablish contact).

Think about it. When you are actually with the person (or thing) that you are attached to, you may not necessarily feel that attachment super powerfully at every moment. You may even take the loved one for granted while you are actually present with him or her. However, as soon as you are separated from the object of your attachment, even briefly, you feel it. The separation that you feel is often described as anxiety. Being separated from whom or what we love makes us anxious. That is unavoidable because it is part of how the attachment works. The effect of attachment is really expressed when we are not with someone, more so than when we are.

Once again, the hormones in our brain that mediate this attachment can actually be measured. This is not conjecture. While we are still filling in the details, much of the brain chemistry that is underneath attachment, love, and grief has been known for some time.[19] The anxiety that we experience as "missing someone" does have a pretty clear biological purpose and an evolutionary benefit: social cohesion. When we talk about social bonding, attachment, and so forth, what we are really talking about is the release of hormones that prime an animal to want to stay with other animal(s) or thing(s). It is like getting "hooked" on being together with our loved ones. In the animal world, these attachment hormones help individuals survive by keeping them with their herds and flocks, as well as their parents, children, mates, and friends. The different kinds of relationships likely involve different precise combinations and doses of hormones, but the attachment works nonetheless. It works because when an animal is separated from her child, parent, mate, or group, the resulting anxiety compels her to reestablish the connection. It drives her to stay in the social group, stick by her mother, care for her child, and so on.

So, if being temporarily separated from someone or something that we are attached to causes us anxiety, if and when the ultimate separation, death, occurs, we get an extreme case of anxiety. This exaggerated experience of anxiety is what we know as grief, and it may indeed be an unavoidable consequence of the prosocial evolution of many animals including and especially humans. There is only one surefire way to avoid grief—never get attached—but that is a rather poor reproductive strategy in a social species.

If the above explanation of grief were true, we would predict that grief would be seen most in species with strong social bonding and least, or not at all, in species with little or no bonding. That is exactly what scientists have found.[20] Similarly, the mechanism of attachment, through hormones like oxytocin, appears universal as well. Of course there are differences across different species, but the theme is the same: attachment-promoting hormones produce a "drive" to stick with the object of the attachment, and the sudden absence of the attachment leads to hormonal anxiety. In a sense, there was really no other way to evolve sociality. Anxiety and grief may be more than just side effects of love and attachment; they may be part of how attachment works.

Once again, the father of the modern study of biology, Charles Darwin, hit the nail right on the head. It only took the scientific community 150 years to fully understand the wisdom of his words. He wrote a letter in 1843 to his cousin, who happened also to be a priest: "Strong affections have always appeared to me, the most noble part of a man's character and the absence of them an irreparable failure; you ought to console yourself with thinking that your grief is the necessary price for having been born with . . . such feelings."

This explanation of grief makes good sense logically, but it leaves me a little unsatisfied. First of all, nature seems better than this—more creative, I mean. If the grief associated with the loss of a loved one is nothing more than a side effect of attachment, it is a very harsh one. It seems to me that Mother Nature has had two or three hundred million years to come up with ways to ameliorate the negative consequences of grief. Instead, such consequences have intensified. As new species have emerged among the social lineages of animals, they have become increasingly more attached and more capable of profound grief.

Alternatively, if Mother Nature cannot dissect the negative consequences from the positive ones for a given evolutionary innovation, she often co-opts the underlying structure for new functions. What I mean is, if nature has not been able to remove grief from attachment, maybe she has provided new functions for grief instead. This way, it is not all bad that animals grieve. In this vein, biologists have suggested a couple of new hypotheses for the social functions of grief.

One such hypothesis is that grieving the loss of a group member allows a reshuffling of the social structure of the group. This is particularly important in species with ranked dominance hierarchies in which reproductive success is closely linked to dominance status. Alpha males and alpha females reproduce more than those further down the pecking order. At the same time, higher-ranking males and females will have more social bonds. They are more "popular," if you will, and will thus be grieved more broadly and intensely than individuals further down the line.

How, then, does grieving promote social harmony? Imagine if an alpha female suddenly dies. One possible result would be a struggle, possibly violent, among viable contenders for the alpha position. However, that is not often what happens, at least not right away. Instead, the group is thrown into a state of grief—a mourning period, if you will. During mourning, animals lose interest in things like dominance struggles, and they temporarily lose any concern for their place in the group. This is a good thing. Without that mourning period, the resulting struggle and chaos could disrupt social harmony in the group and even cause some members to be injured or killed in the fighting.

It is important to keep in mind that the social hierarchy of any group is always in flux. An alpha male or female only reigns for a limited amount of time before he or she is replaced by demotion, death, or expulsion by a stronger contender. Further, there is a pecking order from the alpha down the line to the lowly misfits. However, even the most harmonious social units contain the occasional struggles up and down the ladder as social climbers attempt to claw their way up. In many species, rough-and-tumble play has replaced all-out fighting in these struggles, as discussed in chapter 1. This has preserved healthy competition while removing the survival hazards of true fighting.

Grief, too, could be such a safety measure to keep the peace during a potentially tumultuous time. The tragic or untimely death of a group member, especially a highly ranked one, is a dangerous time for the social order of the group. The resulting grief phase provides a sort of cooling-off period so that the group can slowly ease into the inevitable dominance struggles, so goes the hypothesis.[21] If this were true, we would predict more grieving for higher-ranked individuals and less for lower-ranked ones. This turns out to be the case, offering support to this tentative hypothesis.

Does the same hold for human society? The higher-ranked individuals among us engender more deep and broad mourning than lower-ranked ones. History is filled with stories of tragic deaths of popular leaders, plunging entire peoples into mourning. The streets of America were filled with grief the day that John F. Kennedy was shot, while the homeless and indigent die ungrieved every day. On a more personal scale, imagine how many people would show up to the funeral of a CEO of a medium-sized business who died unexpectedly. Now imagine how many would show at the funeral of a janitor at that same company. Higher rank = more social connections = more grief.

Another hypothesis about the possible biological benefit of grief is that the presence of grief in an animal will elicit sympathy from other animals and allow him or her to form new social bonds to replace the lost one.[22] Strong support for this hypothesis recently came from studies of baboons. Compared with gorillas and chimpanzees, baboons have social standards of conduct that seem more savage and harsh. They fight a lot and can be downright nasty to humans and each other. Nevertheless, baboons appear to grieve and mourn dead comrades.

Researchers from the University of Pennsylvania recently cataloged the social interactions of a troop of wild baboons in Botswana while measuring the levels of circulating glucocorticoids, a group of stress hormones that includes cortisol. They also observed the social and behavioral changes that occurred upon the death of a group member. Not surprisingly, they measured spikes in stress hormones among the baboons when a member died. As expected, the spikes were highest among the baboons that were most closely bonded to the deceased. The most interesting thing that the researchers found was that grieving baboons would often form new social associations and display bonding behaviors, such as grooming and other friendly contact, with members that they had not been closely affiliated with previously.[23]

The point was best made using the story of Sylvia, one of the older female baboons in the group. In general, Sylvia was not particularly friendly. She enjoyed a fairly high rank in the group mostly due to her age and dominance. A cranky old bird, Sylvia was overtly hostile to other females when they would try to approach her. The researchers observing the group had even nicknamed her the "Queen of Mean."[24] However, she was closely bonded to her daughter Sierra. They would groom each other frequently and spend lots of time in friendly contact. This all changed when Sierra was eaten by a lion. (I guess such is life in Botswana.)

Sierra's sudden death plunged Sylvia into a state of depression, and her stress hormone levels soared. She displayed all of the typical grieving behaviors that we have seen throughout this chapter. Most interestingly, she lost her aggression toward the other females in the group. While she previously would hiss and scream at any other female that attempted to be friendly with her, after Sierra's sudden and tragic death, she accepted the approaches of the other females. Sylvia's nasty rebuffs were a thing of the past, and the other females were eager to offer consolation as she mourned the loss of her daughter and best friend. Sylvia gradually emerged from the funk of her depression with eager new grooming partners, and her stress hormones returned to normal. Life marched on, and Sylvia retained her high rank in the group with more friends and allies to boot.

The story of Sylvia is an example of the kind of social recalibration that helps a grieving animal recover and move on. First, the consolation helps to relieve the suffering and pain of the loss and bring the stress hormones back into balance. Second, it helps the bereaved form new friendships to replace the lost one. Researchers had watched Sylvia for fourteen years. Only in the time of her greatest pain did she soften up and form new friendships. Perhaps that is the silver lining that nature has provided us to cope with our grief. In the case of Sylvia, it is conceivable that, if she had remained nasty and hostile after losing Sierra, she would have been in a precarious place in the social order of the group. Without any friends, she may have lost her position and any hope of future reproductive success. Through her grief, she was able to get exactly what she needed: new friends.

To me, this seems similar to what happens when a human suffers great emotional loss. This loss is not limited to death bereavement—other forms of loss often cause us to shift around our social connections. Enduring a divorce, losing a job, declaring bankruptcy, experiencing a serious illness—during these times of pain and loss, we may lose some friendships, strengthen others, and form new ones altogether. You have probably heard the expression, "When times are tough, we discover who our true friends are." The stress and anxiety of grief causes us to reevaluate a lot of our relationships, even unconsciously. When we emerge from our grief, we often move on in life as a "new person" with new social connections and interests. Perhaps the grief itself is what pushes us to explore these social changes as a coping mechanism. If so, well played, Mother Nature.

Earlier in this chapter, we discussed the fact that, apparently, elephants cry. This raises the issue of the evolutionary value of audible and visible grief—crying, wailing, and so forth. Especially for a young animal that has just lost his main source of protection, it would seem that calling out would be the best way to get the attention of the members of the herd. By tugging on their heartstrings, there is hope that others will come to console and provide for the now orphaned child, lest they soon fall victim to nearby predators.

In chapter 3, we saw that social animals are moved by other animals in pain. It stands to reason, then, that emotional pain would be no different. Perhaps by visibly and audibly crying, elephants, dogs, and chimps are really making a desperate attempt for someone to come and offer consolation. I do not mean to say that crying animals are making any conscious efforts, but rather that crying out could be an involuntary expression of grief designed to elicit sympathy. From sympathy comes consolation and attention. From attention comes attachment and social bonding, and—voilà—the orphaned child or now friendless adult has some replacement for his or her loss.

This possible function of visible/audible crying and other biological benefits of grief described here do not necessarily explain how grief evolved in the first place. Those would indeed be very poor attempts to explain where grief came from. Instead, they are possible ways that nature has evolved to "deal with" grief and to use it for some possible advantage, as a silver lining, or to build in some sort of coping mechanisms that will help the animal survive the episode of grief.

For the origin of grief, the only explanation that seems to make any sense is that it is the painful but unavoidable consequence of social attachment.

DO ANIMALS HOLD FUNERALS?

In the summer of 2013, a YouTube video of a dog burying a puppy went viral. In this video, the dog approaches a young puppy that is laying lifeless at the bottom of a large pit in the desert. She briefly smells the dead puppy and then gently uses her nose to push sand on to the puppy. The dog works at this task dutifully for a few minutes until the puppy is well buried. You can search for this video and watch for yourself.

There are many reasons to doubt that this video represents a grieving mother burying her dead child. For one, reports of dogs burying their dead do not abound. Second, this dog is clearly a well-trained working dog, which is easily gleaned from other videos posted by the same user, and even the video in question shows an attention to task that speaks of substantial training. Finally, the video description cannot be easily verified. Without more information, we should be careful not to draw too much from this.

Despite these caveats, the video is haunting to watch. The dog appears withdrawn and forlorn. Her tail does not wag, as would normally be expected when a dog performs a trained task. There is a noted absence of the usual spring in the step that is common in working and performance dogs. Even if this is not her puppy, there is a sense that the dog understands the loss of life at hand. There is no joy in this, only a somber sense of duty.

Do animals have funeral rituals? The very question seems absurd to many. Funerals call to mind codified rites with props, incantations, sacred prayers and anointing, or ritual burial or burning. However, in many species, social grieving events have been documented that include a sort of group mourning and consolation. Some animals have even been seen to partially bury or cover one of their fallen peers.

A clan of African elephants was migrating to new territory when one elderly female collapsed and died. When another elephant noticed, she trumpeted to the others. The entire clan stopped and circled back to the fallen elephant. They took turns poking and prodding the deceased in an apparent attempt to rouse her. Some pulled on her trunk, tusks, and legs in an attempt to lift her to her feet. Some males even briefly mounted her, perhaps attempting to wake her. The group needed to move on to their new location, and so, one by one, they left their fallen friend. However, the next day, many of the group returned—from considerable distance—to check on her one last time. This time, they did not attempt to rouse her. They simply wailed, moaned, and cried, circling her slowly, as if in a viewing procession at a wake.[25]

This was not an isolated incident. In his book, *Elephant Destiny: Biography of an Endangered Species in Africa*, Martin Meredith describes an especially touching event that occurred when the matriarch of an elephant clan passed away:[26]

> The entire family . . . including her young calf, were all gently touching her body with their trunks, trying to lift her. The elephant herd were all rumbling loudly. The calf was observed to be weeping and made sounds that sounded like a scream, but then the entire herd fell incredibly silent. They then began to throw leaves and dirt over the body and broke off tree branches to cover her. They spent the next two days quietly standing over her body. They sometimes had to leave to get water or food, but they would always return.

Many other such instances of elephant death rituals have been observed. There is even a section called Death Ritual in the Wikipedia page on elephant cognition.

Elephants have long been known to recognize and respond to the dead bodies of other elephants, even if they are skeletonized. That is a fairly remarkable feat of memory and cognitive processing. No other animals, not even chimpanzees or dolphins, appear able to recognize skeletal remains of their own species. When elephants come upon the bones of a fallen member of their species, they will pause with immediate recognition. They will sniff the bones extensively, nudge them with their trunks, and often pick them up and carry them away with them. Again, being able to differentiate bones of your own species from those of other species is a surprisingly advanced ability. I would never believe the anecdotes if they were not also accompanied by scientific studies.[27]

Elephants are not the only animals that have a "death ritual." In an episode of a PBS show called *Clever Monkeys*, you can watch how a troop of toque macaques surrounds the body of their dead leader. The various troop members gather around the body in a circle and stare at it in silence. Macaques are not a species that often sit still and quiet. The macaques touch the body of the leader gently, while also stroking, caressing, and grooming each other. Like any pack leader, the dead monkey had rivals and enemies that were constantly nipping at his heels, eager to depose him as the leader. Nevertheless, even those rivals come to pay their respects while he lies dead.

I would be remiss if I did not also mention the death ritual of corvids, a family of birds that includes crows, ravens, and magpies. This will be discussed in detail in the next section.

Is it so farfetched that animals have evolved certain behavioral tendencies in response to the death of a conspecific? After all, funerals are a universal feature of the human response to death. All cultures of humanity have certain rituals that accompany the death of a member of the society. These customs date very far back into human history. Ancient hunter-gatherer cultures were known to build funeral pyres for cremation, ritually bury their dead, or send the bodies downstream in a small boat built especially for that purpose.[28] In all cases, the deceased were adorned in various ways specific to the death ritual.

There is evidence that Neanderthals buried their dead as well, at least on occasion.[29] Earlier claims that Neanderthals included trinkets or rituals in their burials have been strongly disputed, and now most anthropologists believe the burials were purely practical in nature. Nevertheless, burial of the dead may have been a cultural practice that was brewing in the Pleistocene era among Homo sapiens, Neanderthals, and possibly other hominins.

While most modern funeral rituals are dominated by prayers and practices that center around the theme of afterlife, there is some evidence that the central actions of human funerals—burial, cremation, or other means of body removal—actually predate spiritual belief.[30] Initially, body removal was likely more about sanitation and the avoidance of attracting predators, scavengers, vermin, and pestilence. This would have become increasingly more important as human cultures began to settle in one place for longer periods due to the innovations of livestock and agriculture. The shift from nomadic hunting and gathering to farming and stable villages did not happen overnight, and so the need to dispose of decomposing bodies probably inserted itself slowly into human prehistory.

The invention of graves, graveyards, and rudimentary crematoria likely gave rise to the ritualism associated with burial and cremation. It was not long before these inventions and customs merged with developing religious beliefs. Given that grief is the suffering due to the loss of an attachment, it seems rather probable that beliefs about afterlife stemmed, at least in part, from a desire to preserve the lost attachment as a comfort to the bereaved.

According to many psychologists, it is healthy to hold funerals because they function to help the bereaved manage their loss. First, funerals are designed to bring closure by helping those left behind to accept the finality of the person's death and, in so doing, help them move on with their lives. In addition, most experience a surge of grieving during funerals, especially if we have been "bottling it up inside" prior to that. It seems to me that certain aspects of the services are practically designed to ensure that we bawl at least once.

With these two items in mind, it seems plausible that animals might have something to gain from funerals as well. They, too, have a need to accept the permanency of their loss so that they can be open to forming new social connections; and they, too, can benefit from releasing a "grief valve" so as to allow quicker recovery from the spike in stress hormones that are associated with grieving. Thus, it stands to reason that some form of simple funeral ritual might be a good thing for animals.

This notion has not been lost on zookeepers. It has been known for some time that, both in the wild and in captivity, gorillas tend to mourn their dead and "hold vigil" with the dead bodies for a time before leaving them.[31] As with so many other species, the gorillas that were the most closely attached to the deceased are usually the most bereaved. During the collective mourning, members of the group can be seen comforting each other, and stressful cries are common. A grieving band of gorillas in captivity will actively resist the removal of a dead body, becoming agitated and combative, if the handlers attempt to remove the body before the mourning ritual is completed and the gorillas have left the body on their own schedule.

This has led many zoos to hold "wakes" for gorillas when they pass in an attempt to facilitate the grieving process of the rest of the band. In 2004, a female gorilla named Babs died in the veterinary hospital of the Brookfield Zoo outside of Chicago after a long battle with kidney disease. Because she died while away from the gorilla enclosure, the handlers thought that it might help the remaining gorillas reestablish their social hierarchy to know that she had died. They wanted to give them a chance to properly mourn her passing. After they laid Babs's body in the paddock, the gorillas came to view it. Babs's daughter was the first to approach and her mother was the second. As the gorillas sat next to her cold body, they stroked it gently. One gorilla even brought her young daughter, a youngster that had formed a strong bond with Babs, to view the body with her. The young one played with Babs's mouth, just as she did so often when Babs was alive, and laid her head on Babs's chest. Zookeeper Betty Green said: "It was like they used to do in the exhibit, lying side by side on the mountain. [Babs's daughter] rose up and looked at us and moved to Babs's other side, tucked her head under the other arm, and stroked Babs's stomach."[32] As you can imagine, Babs's handlers were moved to tears.

The zoo staff thought that the wake was especially important for this band of gorillas because Babs had been the dominant female. Although the silverback male is the ultimate leader of a gorilla band, the female leaders are far more social than the males. Babs had been the disciplinarian and arbiter of day-to-day operations. If the silverback is the president of the group, the dominant female is the chief of staff, whose daily actions are far more crucial to the social harmony of the band. Babs died well before her time and with no heir apparent. By holding the funeral, the zoo staff hoped to avoid violent clashes to fill the power vacuum. The mood of mourning and mutual comfort helped to soothe the transition.

This was not the first such wake for a gorilla. The Brookfield Zoo got the idea from a zoo in Columbus, Ohio, which in turn got the idea from watching gorillas in the wild. This idea is now spreading to zoos through-out the world, and not just for gorillas. Another example is the story of the funeral for Dorothy, the chimpanzee from a rescue center in Cameroon, which was discussed earlier in this chapter.

GRIEF IN BIRDS

Most of what we can say about grief in animals comes from the study of mammals, especially the big-brained ones: apes, elephants, and cetaceans (whales and dolphins). This is expected for two reasons. First, grief is a result of the loss of something that we are attached to through sociobiological bonding. This kind of hormone-influenced sociality is most pronounced in mammals, and only where there is attachment can there be grief. Second, grief requires more emotional sophistication than most nonmammals show. While some emotional states are so simple that they appear purely instinctual, such as fear, lust, anger, stress, and calm, the more complex emotions require a more developed cerebral cortex and more intricate cognitive processing.

Nevertheless, there are a few hints that a simple form of grief, or something like it, may exist in some nonmammal species, most especially birds. Birds would be the most likely group of animals after mammals to exhibit grief because of their extensive social living and the formation of bonds and attachments.

A study of blue jays published in 2012 had the title, "Western scrub-jay funerals: cacophonous aggregations in response to dead conspecifics."[33] This led to some confusion as the authors did not intend to imply any cognitive or emotional response akin to grief. They were using the term "funeral" only to refer to a gathering event that occurs when a fellow bird dies, and the controversial term appears only in the title, not in the paper itself. Nevertheless, the findings of this study are interesting, particularly because the study describes the results of a systematic series of experiments, rather than just the anecdotes that comprise most of our examples of animal grief.

In this study, the scientists laid out various objects and then watched flocks of blue jays as they discovered the objects. Most objects elicited no response, including pieces of wood painted like blue jays. However, when the scientists arranged a scenario where the blue jays happened upon a dead blue jay, they observed, time and again, a distinct behavioral pattern.

The jays organized into a chorus and sang a song. The other jays seemed to avoid the area for a while, despite the presence of food provided by the researchers.

The study authors suspected that the calls were warnings to fellow jays of danger in the area. This belief was bolstered by another experiment that revealed a similar (but not identical) organized choral reaction when a stuffed owl was introduced. Since owls are regular predators of these jays, this death response may be a sentinel behavior. The blue jays are programmed to respond to the death of another member of their species by warning the others to avoid the area for a while: a predator is about.

Since this is not really an expression of grief, why am I including it here? I find this study thought-provoking because it reveals that the blue jays are capable of identifying a dead member of their own kind. It may seem like simple stimulus-response, but the ability to recognize a dead conspecific is a rather high-level skill for a nonmammal. It means that they have some understanding or at least acknowledgment of what death is. They may not lose sleep thinking about death, but they do know it when they see it.

It turns out that blue jays are not the only birds that have the ability to recognize when one of their own has died. In fact, crows have a so-called death ritual in which members of the flock gather around a dead companion. Vincent Hagel, former president of a local chapter of the Audubon Society, wrote:[34]

> Just a few feet from the house lay an obviously dead crow, and about twelve other crows were hopping in a circle around the body. After a minute or two, one crow flew off for a few seconds, then returned with a small twig or piece of dried grass. It dropped the twig on the body, then flew away. Then, one by one, the other crows each left briefly, one at a time, and returned to drop grass or a twig on the body, then fly off until all were gone, and the body lay alone with twigs lain across it.

This same ritual has been seen in the closely related ravens. In addition, Marc Bekoff recently recorded a slightly different death ritual in magpies:[35]

One approached the corpse, gently pecked at it, just as an elephant would nose the carcass of another elephant, and stepped back. Another magpie did the same thing. Next, one of the magpies flew off, brought back some grass and laid it by the corpse. Another magpie did the same. Then all four stood vigil for a few seconds and one by one flew off.

Apparently, after he published this account, Bekoff was flooded with e-mails from people who had seen this ritual in magpies, as well as in the aforementioned crows and ravens. Magpies, crows, and ravens are all in the same family of birds: the corvids. Guess who else is in the corvid family of birds? Blue jays. While the death-associated behaviors of these corvids may not necessarily indicate grief, it is a social acknowledgment of death, and that is an important first step.

A sanctuary for animals rescued from factory farms reported the tale of Harper and Kohl, two unrelated male ducks rescued from a foie gras production farm.[36] These two ducks both suffered irreparable injuries and disabilities and were very frightened of people. They formed a bond very quickly and clung to each other constantly. This inseparable pair lived for four years on the sanctuary before Kohl could no longer walk or eat. Harper was able to witness the euthanasia of Kohl. Following that, he went to the lifeless body of his friend and quacked frantically, poking and prodding with his bill. After getting nowhere with his attempt to revive his friend, he lay down next to his body, remaining there for several hours.

The handlers at the sanctuary removed Kohl's body and, after a few days, tried to introduce Harper to other potential duck friends. It did not work. Harper spent his time lying in the spots where he and Kohl used to lie, and he was now more afraid of his human handlers than ever. He ate only minimally and avoided contact with all ducks and humans the best he could before finally dying two months after his friend. Did Harper die of grief? What is fair to say is that Harper's health and wellness were negatively impacted by the loss of his only social bond. Already fragile, the loss of this bond almost certainly contributed to his demise.

While most bird species form lifelong pair bonds with social and some-times sexual monogamy (as discussed in chapter 5), geese are perhaps the most prolific of pair bonders. The marital bonds in these species are truly impressive. Any goose hunter can tell you that after shooting down a goose, very frequently one member of the flock will circle back to the fallen mem-ber, despite the terrifying sounds of shots firing, and thus become an easy target. The Nobel Prize–winning scientist Konrad Lorenz spent his life studying these birds and claimed to have only witnessed three instances of a pair-bond dissolution, two of which were "second marriages" that replaced an earlier deceased spouse.[37] Not surprisingly, geese show signs of intense grieving upon the loss of their mates. They will lose weight and withdraw socially. The mortality rate for widows and widowers is very high. The only thing that seems to get them moving and thriving again is if they can form another pair bond quickly, most often with a fellow widow or widower.

This has led to some odd consequences for these geese. Stories abound of lovelorn geese attaching to and following around other non-goose birds and even non-birds. One such Canadian goose, after six weeks of depres-sion and loneliness, became attached to employees of a Dollar General store in Nixa, Missouri.[38]

GRIEF FOR ANIMALS OF OTHER SPECIES

The story of the Canadian goose that ended up attempting to pair bond with employees of a Dollar General brings up the issue of cross-species grief. Animals have been observed to exhibit cross-species empathy, as we dis-cussed in chapter 5, so it makes sense that they may actually display grief for members of other species. In fact, if grief is the emotional response to loss, we would expect animals to grieve when they lose anything they are attached to, regardless of species. That is exactly what we see.

In the summer of 2012, a video appeared on YouTube of a dog mourning the death of her friend, a beaver. The video went viral, articles were writ-ten about the case, and details began to emerge from the owners of the dog. The owners of the dog, named Bella, live out in the country where they keep several dogs with no need for leashes or fences. The beaver, whom they nicknamed Beavis, came around periodically. Since he did not seem to bother the dogs, they did not shoo him away. Bella and Beavis became unlikely friends.

While the other dogs on the ranch were not fans of Beavis, Bella played with him in all the ways that you would expect two dogs to play. They played ball, they wrestled, and they snuggled. Bella's owner described them as "inseparable," and Beavis would frequently seek shelter in Bella's doghouse, where he was welcomed freely. Bella, who protectively guarded her food from the other dogs, would even allow Beavis to eat food out of her bowl.

The now-famous YouTube video captures the aftermath of Beavis's sudden passing. Bella is seen cuddling next to the dead beaver, nosing and licking him, glued to his side. Her cries and whimpers are audible. The other dogs joyfully play in the background in what looks like callous disregard. None of those other dogs were bonded to Beavis; Bella was. Her owners reported that Bella stayed with the body of her dead friend for several hours, periodically laying her head on him and whimpering listlessly the whole time.

Stories of unlikely animal friendships are not hard to find. Animals seem able to readily bond with other animals that they do not feel threatened by, as we discussed in the chapters on love and empathy. Therefore, it makes perfect sense that they would grieve the loss of those bonds. Why not?

One of the most famous tales of interspecies love and loss involves yet another dog named Bella. This Bella, however, had formed a tight bond with an elephant. Yes, an elephant. The fifty- or sixty-pound Bella became inseparable friends with a gigantic, ten-thousand-plus-pound elephant named Tarra. Tarra is the most senior resident of an elephant sanctuary in Tennessee, where she retired after a life spent entertaining humans with her painting skills. (Yes, painting. But I do not want to get distracted from the point here.) Bella also lived in the sanctuary.[39]

Almost immediately after meeting, Tarra and Bella showed affection for each other. Like many dogs, Bella was jittery around animals other than dogs and humans, but she warmed up to Tarra almost immediately. Perhaps because her life had been spent in the exclusive company of non-elephant companions, mostly humans, Tarra showed none of the typical elephant skittishness of small animals. As the unlikely pair spent more and more time together, their mutual trust and attachment grew. Bella would roll onto her back and allow Tarra to gently stroke her stomach with her trunk. The two would swim and splash in the pond together. Tarra would squirt water at Bella from her trunk, much to Bella's delight. Wherever Tarra went, Bella would follow. The two were free to roam the grounds of the sanctuary day and night, and happily did so at each other's side.

One night, Bella was attacked and killed, most likely by coyotes. While coyotes would never have dared to approach Tarra, Bella was known to run off from Tarra for short periods, especially when Tarra was sleeping. That seems the likely scenario behind this tragedy.

The most haunting part of this story is that, from the condition and location of Bella's body, it became clear to those investigating that the attack could not have happened where she was found, nor could she have walked to the location, as the injuries would have been fatal very quickly. Somehow, Bella was moved some distance after she died. Someone brought her broken body back to the home base. The staff went to Tarra and found blood on her trunk. No injuries, just blood. It seems that, after discovering it, Tarra had carried the bloodied body of her friend Bella back to where the sanctuary staff would find it.

Although Tarra eventually recovered from her loss, she was initially plagued by the typical signs of elephant grieving. She avoided social interactions with the staff and other elephants and was withdrawn from activity in general. Normally, she was among the most curious elephants, always exploring new things and poking her trunk into everyone's business. She lost that enthusiasm for a while, but the staff of the sanctuary reports that she has now regained the spring in her step and is forming new social bonds. You can read more about Tarra and Bella at www.elephants.com.

It is probably true that dogs are a bit more "susceptible" to cross-species bonding and grieving since they have been bred and heavily selected for their companionship with humans. Thus, it is not surprising that dogs would form bonds with animals like beavers and elephants. My own dog, Bruno, is very strongly attached to my better half and me. If either one of us is traveling or stays out late, the remaining person is "not enough" for Bruno when it is time for sleeping. Bruno sleeps in bed with us, usually right in between, and when one of us is not there, he just cannot seem to relax and go to sleep. He's usually a very lazy dog that loves bedtime snuggling and sleeping, but when the pack is not all there together, he is grieved. He will sit up at the foot of the bed and stare at the door, whining and whimpering periodically. It is a heartbreaking scene.

Nevertheless, it is not just dogs that are capable of bonding with members of other species. Koko, probably the most famous gorilla of all, was known to form attachments to her human handlers as well as other animals around her. Koko was most noted for her remarkable use of language, which we will discuss later, but a story that is not told as often is her tendency to adopt pets. In fact, in 1984, Koko signed that she would like a cat for Christmas. Not sure that this was a good idea, the staff of the Gorilla Foundation gave her a stuffed animal cat instead. This did not at all satisfy her, and she repeatedly signed that she was sad.

After she repeated her request, the staff decided to give it a try and arranged to give her a kitten for her birthday in July of the next year. They allowed Koko to pick out the cat from a litter of abandoned Manx kittens. Manx is an odd cat breed for its lack of tail. Koko selected a gray male and named him All Ball. As told in the book *Koko's Kitten*, Koko was very dedicated and attached to the kitten.[40] She was remarkably gentle and tender with him. She would carry All Ball everywhere, and she even attempted to nurse him in a manner identical to how wild gorillas nurse their young.

Why would she attempt to nurse him? Well, as we know, oxytocin gets released during social bonding. Oxytocin is also the hormone most involved in nursing. It is possible that, given Koko's very unique and peculiar social setting, surrounded by humans indoors instead of gorillas in the jungle, the social parts of her brain were wired up a little differently. Maybe the oxytocin release was exaggerated or simply misfired and tricked her into feeling as though she were a new mother. Maybe she was play-mothering like the vervet monkeys discussed in chapter 1. Who knows? In any event, Koko was very attached to her new pet and cared for him gently and dutifully.

Because we are in the chapter on grief, you may already be braced for where this is going. At some point, All Ball ventured out of Koko's cage, out of the facility, and was struck and killed by a car. The researchers told Koko that All Ball had gone away. Koko became very distraught and agitated. She continually called for All Ball with signs and vocalization. She signed to her human handlers, *sad, bad, sad, bad* in rapid succession. This gradually shifted to the signs for *frown, sad,* and *cry.* Later that night, Koko made sounds that the handlers had not heard her make before. Her owner and chief handler, Dr. Penny Patterson, described it as weeping.

The next year, Koko was permitted to pick out a new pet. Although her options included several breeds this time, she chose two tailless Manx cats, like All Ball was. She named her new pets Lipstick and Smokey. Koko doted on these kittens the same way she had doted on All Ball and, thankfully, was able to enjoy them for a lot longer.

Pets also grieve for their owners. The Internet is filled with tales of pets, mostly dogs, grieving the deaths of their human companions. There are pictures of dogs clinging to lifeless bodies, sitting underneath coffins, and lying on top of graves. Usually, the details and backstories are scanty, but the stories are too numerous to dismiss.

* * *

In summary, I think it is safe to say that animals display many of the same "symptoms" of grief that humans do. They withdraw socially, eat less, and play less. It even appears as if some animals experience a bit of denial. Denial is often the first stage of grief in humans. It is not that someone truly believes that the deceased is still alive. It is more a temporary "suspension" of belief, a defense mechanism because the reality of the death is too painful to acknowledge just yet. Denial is what came to mind when I read about the many kinds of animals that carry around their dead children for a while before letting them go.

Social attachments are a feature of the lifestyle of many animals, including almost all mammals. These attachments work, in part, by drawing us back to our companions when we are separated from them. We experience discomfort in their absence as a means to drive us to find them and reunite. This is all controlled by hormonal signaling in the brain, and experiments with animals have shown that. The neurochemistry underpinning grief in humans is, as far as we can tell, identical to that of other mammals. Attachments are formed and strengthened through hormones, and the withdrawal of those social inputs leads to an imbalanced state. The imbalance, if slight and temporary, leads to some mild anxiety. This is "missing someone." If the withdrawal of the social attachment is permanent, however, the imbalance is much worse. Anxiety spikes, and we call that grief.

As mentioned earlier, grief is most likely a side effect of attachment. It is probably unavoidable. If we did not miss people and grieve their loss, the attachment itself would not really work. What is more interesting, I think, is that mammals have adapted to the curse of grief by finding silver linings. As mentioned above, grief brings about changes in social behavior that will actually help us recover from our grief by forming new attachments. That is just as true in animals as it is in humans.

Despite this, some stubborn people may claim that animals are only acting as if they are grieving. Words like "love," "sadness," and "grief" seem to irk people when applied to animals. To this, I can offer a compromise. If we substitute the words "social attachment" for "love" and "attachment withdrawal" for "grief," then I think we would describe the reaction to the death of a mate in almost exactly the same way for elephants as we would for humans, save for descriptions of the inner emotional feelings. Further, the only reason we have to make an exception for the descriptions of emotional feelings is because we do not have access to them, not because we have any evidence that they are different. Further still, in the few animals that can communicate their feelings, what do they tell us? Koko feverishly made the signs for *sad*, *cry*, and *hurt* when her cat died. How else can that be interpreted?

My point here is that humans and animals do the same things when we experience loss, so it is likely that we feel similar things as well. I think that should be the default position, and as such, I will give the last word to Barbara King: "Where there is grief, there was love."

FURTHER READING

Bekoff, M. "Grieving Animals: Saying Goodbye to Friends and Family." *Psychology Today*, https://www.psychologytoday.com/blog/animal-emotions/201207/grieving-animals -saying-goodbye-friends-and-family.

King, B. J. "When Animals Mourn." *Scientific American* 309 (2013): 62–67.

King, Barbara J. *How Animals Grieve*. Chicago: University of Chicago Press, 2013.

Masson, Jeffrey M. *When Elephants Weep: The Emotional Lives of Animals*. New York: Delta, 1995.

7

JEALOUS BEASTS

The Dark Side of Love

THE DISCUSSION OF jealousy begins a new and depressing unit in this book. Thus far, we have been talking about how kind and loving animals can be. We have talked about chimpanzees that took care of their developmentally disabled friend, a dog that grieved for a beaver, an elephant that refused to harm a dog, Canada geese that mate for life, dolphins that refuse to let go of their dead children, penguins that adopt abandoned children, wolves that insist on playing by the rules, and baboons that comforted a lonely mother who lost her child. We have been able to see the best of ourselves within these animals. This has been a heartwarming discussion of the virtues of fair, generous, and loving animals that parallel our own fairness, generosity, and love.

Now, we must talk about our dark side. Although we have endless capacity to do good and to love others selflessly, we also have a capacity for great evil. Ruthlessness can take many forms, but the unifying theme is total disregard for the well-being of others in order to single-mindedly further one's own interests. Examples of this fill our history books and current events. The same goes for animal life in the wild. There is no shortage of ruthlessness in nature.

In the previous chapters, I have made a strong argument that the evolution of many animal lineages, including humans, is marked by cooperation, reciprocation, and social attachments. I made this argument to contrast the traditional view of natural selection as working solely through intense and ruthless competition. I have downplayed the competitive side of animal life because that has received enough attention. The truth is that both views are correct. The evolution of social animals has been a balancing act between cooperation and competition, and some species are more competitive while others are more cooperative.

For example, among our closest relatives, chimpanzees and bonobos are much more cooperative and social than are orangutans. While chimpanzee life can certainly involve aggression and fighting, it is also marked by peaceful dispute resolution, reciprocal altruism and generosity, and affection. Most chimpanzee fighting, in fact, is between-group hostility, not within-group. Orangutans, on the other hand, are solitary as adults and generally hostile to each other. The only real bonding that takes place is between mothers and their children, but even this occurs only when they are young and dependent. Mothers and their adult children show no special connection, affection, or cooperation. In fact, the impressive intelligence of the orangutan is a major challenge to the prevailing hypothesis that sociality is the key driving force for the evolution of higher cognitive abilities.[1]

Just as each species is different, individuals within a species can vary as well. Of course, we know that some humans are much more generous and compassionate than others, and the same is true for animals. In any given species, there will be great variability in their virtuousness, so to speak. Each individual has his or her own center of gravity with regard to competition and cooperation, which can even shift over time based on the reality on the ground. Natural selection works to favor one strategy over the other when the environmental conditions call for it. Meanwhile, mutation and sexual recombination keep supplying the diversity, so we always have some variability in just about all traits. Nature always stands ready to send a gentle species careening toward its darker instincts, should conditions require it.

As much as I wish this book could be nothing but rainbows and gumdrops, it is time to leave the comfort of the prosocial side of human nature and venture into the darkness of our antisocial side. Each one of us has our demons, as well as our better angels, because we are all a product of evolutionary forces that favored generosity in some instances and selfishness in others. Maybe by understanding our demons, we can disarm them.

DEFINING JEALOUSY

Jealousy is one of those human emotions that is not always easy to define without venturing into other emotional states like envy, bitterness, and even resentment. Yet we all know what jealousy is, even without having a handy definition. We know it, because we have felt it.

Merriam-Webster gives three definitions of jealousy:

1. (a) the intolerance of rivalry or unfaithfulness; (b) the disposition to suspect rivalry or unfaithfulness
2. hostility toward a rival or one believed to enjoy an advantage
3. vigilance in guarding a possession

The first two definitions seem to focus on people and relationships, while the third definition focuses on things. For the purpose of this chapter, I will stick to the first two definitions, those having to do with rivalry and unfaithfulness. The third use of jealousy, in my view, ventures into the territory of greed and envy, which will be covered in the next chapter. Having just finished the chapters on love, attachment, and grief, I think it is better to discuss jealousy as it pertains to personal relationships before we get into the issue of possessions.

I should also mention that the distinction between jealousy and envy has some regional differences. For example, it seems that in the United Kingdom, the distinction between jealousy and envy has more to do with whether you already have the thing or relationship in question. The definitions in the Oxford dictionary indicate that jealousy is the fear of losing people or things that you already have, while envy is coveting things that you do not have. I am not quarreling with the precise uses of the words "jealousy" and "envy." After all, who am I to tell the English how to speak English? All I am saying is that in this chapter I will focus on the kind of jealousy that deals with relationships and individuals, not things.

For our purposes, jealousy encompasses at least three feelings: (1) the fear of losing a relationship, (2) the negative reaction we feel when a rival threatens to disrupt a relationship (or merely when we perceive such a threat), and (3) the continued anger and/or despair when a relationship is lost to a rival. Further still, the relationship under threat need not be sexual/intimate. We can be jealous of the love that a parent gives another sibling. We can be jealous of the attention that our best friend gives to a new friend. We can be jealous when our boss showers praise on a colleague. And, of course, we can be jealous when our spouse looks longingly at another. Jealousy is all of that and more.

SEXUAL JEALOUSY AND MATE GUARDING IN ANIMALS

The kind of jealousy that we probably think of first is also the easiest to understand biologically. In humans, sexual jealousy is the fear that an intimate partner is not being sexually faithful. Romantic jealousy is the fear that an intimate partner may fall in love with another and thus end the relationship in favor of a new one. These two can sometimes be distinguished from one another, just as sex and love can sometimes be separated. For example, some people tolerate some extramarital sexual activity by their spouse but would be severely threatened by the possibility of a spouse leaving the marriage altogether. Also, in open marriages, sexual monogamy is never expected, but the love and commitment of the marriage is generally understood as exclusive, and thus, romantic jealousy can still appear.

For simplicity's sake, I will discuss the issue of sexual and romantic jealousy as one single topic. Most human marriages and intimate relationships operate that way, and I am not sure that they can be cleanly dissected from one another in animals, either. Furthermore, with humans, it is difficult to say what role culture has played in forming our understandings of faithfulness and jealousy, which may obscure the underlying biology. We are still not really sure what the underlying natural biological state of the human family really is or if there even is such a thing as a natural state. There is plenty of evidence that early human families were extended clans with communal parenting and little sexual monogamy. This is certainly how most of our primate cousins build families.

On the other hand, sexually exclusive binary marriage, though certainly not universal, is the family paradigm that is the most widespread in native cultures throughout the world. Even in those traditions and cultures that espouse plural marriage, sex outside of marriage is considered prohibited and would evoke jealousy from the cuckolded spouse. Meanwhile, still today in some Amazonian tribes, sexual fidelity is unheard of and ritualistic group sex occurs. And yet, families consist of male-female marriages and child-rearing. The point here is that it is very difficult to differentiate cultural underpinnings from biological ones when it comes to sexual and romantic jealousy. Our psychology is the result of both our biology and our social experience. I think the study of jealousy in nonhuman animals will help us sort that out.

In 2011, a pair of young Malayan tigers named Seri (female) and Wzui (male) were brought to the El Paso Zoo in the hopes that Wzui would impregnate Seri, as well as another female already living at the zoo, Meli, who had lost her mate to cancer. Malayan tigers are highly endangered, and zoos have begun breeding programs in which they periodically exchange tigers in an effort to promote genetic diversity. Little is known about the mating behavior of this species, and breeding them in captivity has proven difficult, given the very narrow fertile period of just three days each year.

Seri and Wzui appeared to form a mated pair very quickly with plenty of sex and social bonding. Meli, who was much older and had been at the zoo for ten years already, ignored the young lovebirds for the most part. However, after some months, the zookeepers noticed that Meli and Wzui began to "flirt" through the fence that separated them—that is, they began paying attention to one another and made the purring sound that is widely recognized as a sign of affection and friendly greeting among felines. This enraged Seri, and she began to make aggressive displays toward Meli, who responded in kind. As the animosity between the two females grew stronger, Wzui's interest in Meli continued to develop, even though they were being kept in separate areas and could not touch each other. The zoo even put out a press release reporting that a tense "love triangle" had developed among the tigers.[2]

One day, after an affectionate morning of sex and grooming between Seri and Wzui, Wzui was spotted, once again, looking at Meli. Meli flirted back, and the two watched each other through the fence for a while. This was the last straw for Seri. In a surprise attack, she lunged at her mate, going directly for his throat. Wzui was killed almost instantly.

What else can we say but that Wzui was the victim of jealous rage? Seri wanted him all to herself, and if he would not stay faithful, he would pay the price. If Meli had not been in a separate pen, Meli and Seri probably would have fought openly. Frustrated that she could not attack the real target, Seri focused on the target that she could reach.

An even more bizarre incident of jealous rage in animals dates back to 1902 in a zoo in Marseilles, France. It seemed that the lone gorilla at the zoo, a male silverback named François, had become quite taken with his human handler, Journoux. It is not clear if this attachment was one of a friendly nature or of sexual lust. Although it seems odd, it could have indeed been sexual/romantic, given how long this silverback had been without the company of other gorillas. In any event, Journoux got married and brought his wife to see the zoo animals under his care. François was immediately uneasy. Somehow, he figured out that the relationship between Journoux and his wife was different than that of Journoux's other relationships. This one was a threat to him. François became hostile to the interloping wife and also to Journoux himself. This went on for some time before Journoux attempted to resolve the conflict by approaching François alone and offering him comfort and consolation one-on-one. This well-meaning attempt ended with the violent deaths of both François and Journoux, each at the hand of the other.[3]

Tigers and gorillas are not the only animals in which we see outbursts of jealous rage. There is a species of marine crustacean called cleaner shrimp. These little guys are fiercely territorial and form pair bonds for mating that are remarkably long-lasting and monogamous, something that is almost unheard of for invertebrates. The pair bonding is enforced quite viciously.[4] If intruders approach, the jealous mates chase them away, and if they catch the would-be interlopers, they often kill them. They have also been observed to kill their mates in response to infidelity.[5] And now for the really weird part: these shrimp are all hermaphrodites. If you think we mammals have it bad, a hermaphrodite must watch its mate vigilantly because every single other member of the species is a potential rival.

It is sort of amazing that a fully hermaphroditic species evolved into a social structure that enforces monogamy. One would imagine that none more than hermaphrodites would tend to favor as large a number of mates as possible due to their need to promote genetic diversity. Also, there cannot be gendered behaviors among hermaphrodites, since there is only one sex. But there you have it: sexual jealousy among hermaphroditic shrimp.

Now a look at our fellow primates: The titi monkey, a little-known species of New World monkey related to capuchins, tamarins, squirrel monkeys, and howler monkeys, has a social organization that closely resembles the most common family structure found in present human culture. Male-female pairs mate for life, and the "pack" for these monkeys consists of the nuclear family only: parents and children. When the adolescents become fully mature, they leave their families, find mates, and start families of their own.[6]

Researchers at the California Primate Research Center performed a series of experiments with titis through which they probed the social behaviors surrounding monogamy. All of these studies revealed patterns of behavior that are eerily similar to those of human monogamy. One in particular probed for jealousy. The scientists took monogamous pairs and introduced potential sex rivals, placing them at varying degrees of proximity.[7] The researchers then observed the reactions of the mated couple. Interestingly, the researchers found that the "married men" reacted to the approach of an intruder male by getting closer and closer to their "wives" and by displaying increasing amounts of aggression to the intruder. In other words, they acted a little jealous. Unmated males were largely indifferent toward intruder males.

Many baboon species have a social structure in which a troop consists of several males and several females, each with separate dominance hierarchies.[8] Most of the time, sexual activity is common among all individuals in all combinations. However, things get stricter when the females are in estrus and males' access to females becomes restricted. The females, in order of their dominance ranking, choose the males. Although females are born into their troops and never leave, males will roam about every few years, particularly when they are bumped from their places by young social climbers.

Once a male baboon has been chosen by a female and has mated with her, he will guard her jealously for the rest of her estrus. He will follow her around, restrict her movements, and become hyper-aggressive toward any male that approaches. He will also mate with her repeatedly over this time period. (You know, just in case.) So busy are the male baboons with mate guarding that, during the estrus phase, they will lose sleep and miss meals. They lose all trust in other members of the troop, including and especially their mates, and also including brothers and lifelong male friends. Trust and friendship are temporarily replaced with suspicion and jealousy.[9]

The interesting thing about these baboons is that all goes back to normal when estrus is over. No more mate guarding; no more aggression and hostility; no more jealousy. Females can go back to having sex whenever and with whomever they want. As such, jealousy is focused pretty clearly on the matter of procreative sex. The males protect their reproductive investment, pure and simple. By guarding their female mate, they ensure that the child that results will truly be theirs.

Paternal investment in offspring is an important consideration in baboons because, in most cases, offspring gain the protection of their father as long as he is certain of his paternity and remains in a relatively dominant position. On the other hand, when a new alpha male takes over, he sometimes kills all the small babies in the troop.[10] He does this not only to eliminate competition for his own children but also in order to bring the nursing mothers into estrus again, so that they can direct their reproductive and maternal energy to his children, not their future rivals.

The mate guarding that occurs in baboons is common in other primates as well. In gorillas, the fertile period of the females is not as limited as in baboons. Probably for that reason, gorilla troops only consist of one adult male: the silverback. Somewhere in gorilla evolution, the extended female fertility and constant male mate guarding made it impossible for males to live together, and the harem social structure evolved as the only harmonious scenario. When one silverback is successfully deposed by an invading male, what do you suppose happens to the youngsters? You guessed it—he kills them all.[11]

The London Zoo learned that lesson the hard way.[12] Their gorilla enclosure contained a small band of gorillas organized into a harem. In 2010, the silverback died suddenly from diabetes. As expected, the females became very agitated and anxious. The harem is part of their natural social state, and deviations from that cause great stress and a resulting decline in health. Females in a harem without a silverback will even pull their hair out. Making matters worse, one of the females at the London Zoo was about to give birth.

The zookeepers were faced with a terrible choice. If they were to introduce a new silverback, the infant gorilla, once born, could be killed by him. If they did not introduce one, the condition of the three adult females would continue to deteriorate and the infant could die from malnourishment anyway. He could even get caught in the crossfire between the agitated females. (Adult female gorillas are generally hostile to each other; their fragile peace is kept by the silverback.) The zoo chose to introduce a new silverback, Kesho, and resolved to monitor the situation carefully. Soon after, an adorable young male named Tiny was born.

To their credit, the zoo staff handpicked Kesho from all the silverbacks they had access to. They saw him as the most likely to make the transition peacefully. He was only eleven years old, well before his testosterone-peaking years. He was untried as a leader and somewhat submissive and bashful, hardly the brutish bully that we picture with silverbacks. The zoo staff kept the pregnant female in a separate enclosure and waited until

Tiny was born healthy and Kesho had bonded with the mother through the fence to slowly begin introducing them. In short, the zoo staff had done everything they could. Besides, who could possibly harm a cuddly and innocent infant? It turns out that Kesho could. He savagely beat Tiny to death during their second closely supervised visit.

This is a tragic lesson that silverbacks are not at all interested in adoptive fatherhood. They know which children are their biological offspring, and those that are not are in grave danger. A male gorilla cannot waste the precious parental resources of his troop to bring up the offspring of other males. There is just no room for that kind of generosity in jungle life.

There is another feature of silverbacks that is a natural consequence of the harem social structure: they are unabashedly misogynistic. Silverbacks are well known to dote on their sons but not so much their daughters. After their sons have weaned, a silverback takes over the feeding responsibilities for the young ones. When he does so, he will ensure that all of his little sons get their fill before he offers any to his daughters. These hyper-alpha males are fiercely protective of their sons and do not appear to give much regard to their daughters.[13] Come to think of it, I know some hyper-alpha male humans like this.

For a silverback, the ultimate priority is running the harem in a way that promotes the success of his genetic offspring, and children of other males cannot be tolerated. They favor their sons over their daughters because their sons will need to grow up big and strong in order to fight for the privilege to run a harem. Life for the young males will be difficult—few males survive to run their own harems. On the other hand, the silverbacks know that their daughters will be just fine. Female gorillas already have a good shot at passing on their genes without much fuss. The silverback does not need to dote on his daughters; it is the sons who are in a precarious position. There are always several females for every one male. Where are all the leftover males? Killed by rival males.

What does this say about jealousy? The gorilla harem is the ultimate expression of mate guarding. One male completely dominates several females to the exclusion of all other males. Any possible intruder male will be viciously attacked, and any children of the previous harem leader will be killed. This is sexual jealousy taken to its logical extreme. Thankfully, most human males do not commit jealous murder in their lifetimes. However, I bet most men have contemplated it, however briefly and unrealistically. Even if most would never do it, they would think about it—even fantasize about it. And what is fantasy if not entertaining the notions brought into our consciousness by instinct? Our "natural state" might be more murderous than we think.

SEXUAL AND ROMANTIC JEALOUSY IN HUMANS

Is there reason to believe that human jealousy is akin to the mate guarding that we see so commonly in nature? The picture is muddy because of the strong influence of cultural norms on gendered behaviors. The traditional view that, in animals, "males are promiscuous; females are coy" dates back to Darwin himself. This was likely a projection of Victorian sensibilities regarding human sexuality onto animals and is the one area in which Darwin very likely mucked things up. Earlier generations of psychologists and biologists operated under the assumption that men were more likely to engage in infidelity and become sexually jealous.[14] As they say, a thief thinks everyone steals. However, more modern studies have revealed that sex differences when it comes to love, fidelity, and attachment are not as great as once believed. The pall of Victorian values on animal sexuality is beginning to lift.

In contrast to what was believed about the sexual jealousy of men, it was once believed that women were more likely to experience emotional jealousy toward their male partners. It was thought that women are more scared of their husbands enjoying the company of others—not necessarily other women—to theirs. This evokes a rather 1950s style of sexual politics. As I said, the gap between the sexes is narrowing. This could be because of the erosion of gender bias among scientists, or it could be that the culture really is changing right underneath us. I suspect both. The world is constantly changing. Both we, the subjects of study, and we, the scientists doing the studying, have changed in all ways related to gendered behaviors. Those changes are still underway, but we will proceed and explore what we know (or think we know) about the biology.

In species that engage in the harem lifestyle of one male dominating the sexual reproduction of multiple females, such as lions and gorillas, there tends to be large size differences between the sexes. Male gorillas are typically twice the size of female ones, for example. This is because dominance of a harem is typically won by fighting other males, often to the death. This violence-based system does not much hurt the overall reproductive potential of the species because females are the limiting factor in reproduction and the fighting rarely harms them. On the other hand, in species that are monogamous, males and females are typically the same size. Gibbons are a good primate example of this. Similarly, in species in which sexual freedom and promiscuity is so lush that sexual rivalry is nonexistent (like bonobos), size differences between the sexes also appear negligible.

Human males are, on average, larger than females. Does this mean that we are evolutionarily adapted to the harem lifestyle? I suspect not. First of all, the sex-size difference in humans is nowhere near as stark as it is in gorillas or lions. Furthermore, some human ancestors showed much greater size differences than we do. In other words, in our recent evolutionary lineage, we seem to have evolved away from size differences between the sexes. Even if the harem lifestyle was once present in our distant ancestors, all anthropological and archaeological evidence indicates that our species moved away from it long ago.

However, the question remains: are we wired for paired monogamy, as suggested by modern and historical culture, or for communal social units, as suggested by evidence of prehistoric human society? Although likely unanswerable, this is an important question because mate guarding could either be something that humans have been evolving away from or something that we have been evolving toward. Remember that the development and evolution of behaviors takes time. Any particular moment in time, including the present moment, is only a snapshot of a system that is in great flux with no particular target end point.

Further still, the historical male sexual dominance over women that, until recently, was widespread in human cultures is not the norm in other primates.[15] Once again, it is hard to say if the historic human tendency for males to dominate the sexuality of females stems from biological factors or sociological ones. One thing is for certain: the gradual progression away from that cultural phenomenon is bringing relief and freedom to women. While we still have a ways to go, especially in some parts of the world, this liberation could rightly be described as an evolution—at least a cultural one.

We can ponder whether humans are built for monogamy until the cows come home (and I suspect we will), but there is no mistaking that humans are indeed prone to sexual and romantic jealousy. While I certainly do not quarrel with the notion that sociological factors play a large role in how this jealousy takes shape, I do think there is strong reason to believe that biology underpins some of it. When you were reading about the sexual jealousy of animals earlier in this chapter, did it sound very familiar?

Think back to the episode of Seri, the tigress who killed her husband with the wandering eye, or the hermaphroditic cleaner shrimp that viciously attack and kill any approaching rival. While these episodes may appear savage and brutal, are they so different from what happens when jealous humans lose control?

Jealous killings happen so often that there is a term for it: a "crime of passion," although technically any rage-associated killing could also be classified as such. According to the U.S. Bureau of Justice Statistics, 30 percent of all female murder victims were killed by their husbands and nearly 20 percent were killed by their ex-husbands.[16] Even when jealous killing does not actually happen, it almost does. There was a study of five thousand people, both women and men, from six different cultures, in which 84 percent of women and 91 percent of men admitted to having at least one fantasy about killing their lover or a romantic rival.[17] Those are big numbers. This is why, when a married person is killed mysteriously, any detective will tell you that the first person the police need to rule out is the spouse. Of all the seven billion people on the planet, the one person that you are most likely to be killed by is your spouse. Sleep tight.

Furthermore, when it comes to jealously protecting and promoting paternity, humans are not that different than gorillas. You probably recoiled in horror a few pages ago when you read about Kesho viciously killing the infant Tiny. What Kesho did was terrible. Awful. Horrific. It was also perfectly in line with what just about every human civilization did when they conquered other civilizations. Whether the Babylonians, Mongolians, Greeks, Romans, Mayans, Visigoths, or Chinese, a pretty universal phenomenon of conquering peoples was to kill the men and children of the vanquished people and take the women as wives or concubines. This was part of how an empire integrated new people so quickly. It was not just cultural assimilation; it was genetic assimilation, facilitated by the murder of the men and their offspring and the rape and forced marriage of the women.[18]

You do not have to reference prehistory, classic antiquity, or contemporary hunter-gatherer tribes to find this barbaric behavior. Christopher Columbus "discovered" the New World just over five hundred years ago. Following this watershed historic moment came the systematic eradication and enslavement of the native peoples and cultures. However, they did not suffer genetic extinction. The entire ethnicity we know as mestizo, which includes the majority of Latinos, is largely the result of Spanish men fathering children with Native American women. Mestizos comprise the largest racial group in the Americas and one of the largest on the planet. This is just a recent example of the human predilection toward reproductive dominance.

When it comes to jealousy, humans really are not any more evolved than primates, and if the evidence here does not convince you, maybe some brain scans will. Scientists at Emory University have studied the brain activity of jealous monkeys using the technique of Positron Emission Tomography (PET) scans.[19] Specifically, they monitored the brains of male rhesus macaques as they watched their female mates. Then, they introduced a male rival and observed how brain activity changed. PET scanning exposed which parts of the brain were active at a specific moment in time. This research revealed a neural pathway that corresponds with jealousy—the "jealousy program," if you will.

Using this same technique, another research group went a step further and explored how the jealousy program operates in a monogamous species of primate, the titi monkey.[20] Using these monkeys, the scientists were able to look for the jealousy program before and after pair bonding. Sure enough, the jealousy program only gets "installed" in the male titi monkey after he has formed a pair bond and is activated when a rival approaches his mate, not another female.

This study builds on the earlier work in macaques in two important ways. First, it clarifies the fact that pair bonding is what creates the jealousy program and that mate guarding is what activates it. Second, the examination of titi monkey brain activity shows that the jealousy software runs in the primate brain in similar ways in monogamous and nonmonogamous species. This is key. In my view, romantic jealousy is unique to monogamy, while sexual jealousy can be present in any sexual species. That the jealousy program is so similar in monogamous and promiscuous species argues that romantic and sexual jealousy are really the same response, or at least extremely similar ones. In terms of brain activity, there does not appear to be any distinction.

Since romantic and sexual jealousy may be the same thing, the distinction between them in humans is found in what "trips" the jealousy program to become active. For romantic jealousy, the activating event is a threat (real or perceived) to the emotional attachment; for sexual jealousy, the stimulus is a threat to the sexual monogamy. However, once activated, the jealousy software runs on the brain hardware in the same manner and produces the same results: protectiveness, guarding, anger, despair, and potentially violence.

To demonstrate this, scientists ran the jealousy experiments on humans. Although this study focused largely on discovering differences between men and women, scientists in Japan monitored the jealousy reaction in the human brain using PET scanning and found that it is remarkably similar to that of macaques and titi monkeys.[21] What this means is that the jealousy program is ancient and hardwired into the human brain. Of course, various species will evolve differences in how the program is activated and what the output will be, but the program itself is shared.

Also, I should say that I do not dispute the role of social conditioning and culture on our experience of jealousy. Some people are clearly primed for dramatic jealous responses, while others seem to be oblivious of blatant infidelity, practically right in front of their eyes. I doubt that those differences are genetic. My point here is to say that jealousy is an innate feature of humanity and something that we share with other mammals.

THE ROOTS OF INFIDELITY
AND SEXUAL JEALOUSY

Even for those that contend that binary parenting and the traditional nuclear family are the natural state for Homo sapiens, there is good reason to believe that a drive toward infidelity remains present. If that traditional view of marriage and family life were biologically correct, the evolutionary basis of the drive toward infidelity would be different, but similar, between the sexes. For an early human man, infidelity would simply be a way to achieve additional paternity. By sowing his oats in as many fields as possible, he has a greater chance of leaving offspring, and thus his genes for infidelity, in great abundance. The genes that underpin the drive to procreate freely and widely would tend to accumulate. It is simple math.

In early human women, however, the drive toward extramarital sex was probably a bit different. Unlike birds, they were not able to simply sleep around the village and drop their eggs off for others to rear and raise. Nest parasitism is not really possible for mammals. Any additional maternity that a female achieves through infidelity, she must also care for, at least during the gestation and nursing phases. As long as she has a mate, why would there be any reason to look for extramarital sex at all? She already has access to the means of producing offspring as fast as she can. She, not the males around her, is the limiting factor in her reproductive capacity. This simple logic has, in part, led to the Victorian belief that still pervades human culture that women have far less tendency toward infidelity than men. The problem is that this logic gets lots of the biology wrong.

Women do indeed have several reasons to want to engage in infidelity. First of all, while females of all species do their very best to select the best mate that they can, we all know that different individuals have different strengths. There is no one perfect man (or woman). As females (and males!) begin producing their offspring, there is a strong incentive to mate with several suitable mates in order to produce sons and daughters with great variety in their genetic toolkit. The best bet in ensuring the success of progenitors is not just by having lots of them but by having lots of *kinds* of them: tall, short, fast, strong, camouflaged, brightly colored, good climbers, fast runners, quick thinkers, good instincts, and so on. The best way for a female to achieve this diversity in her offspring is to mate with lots of different males, not just the one she may be life paired with. For her, it is not about quantity; it is about quality and diversity.[22]

There are other incentives that primate females have for engaging in infidelity. First, by mating with other males, females help to protect their offspring in the event that one of her lovers moves up the dominance hierarchy. As we have already discussed, when a new male takes over a harem (as in gorillas) or even when a new alpha becomes the head of a multi-male troop (as in baboons and chimpanzees), he often kills the previous alpha male and possibly some potential rivals, as well as any children from the previous alpha. However, a new alpha may show mercy to the children of a female that he has repeatedly mated with. After all, some of those children may actually be his. Because of this phenomenon, female infidelity is something of an insurance policy to protect one's children from future harm when a new dominant male takes over.

Modern humans like to think that we are above all of this. So be it. My point here is only to say that there are plenty of explanations for intrinsic instincts toward extramarital sex in both sexes. Recent surveys have borne this out, showing that the incidence of, and temptation toward, marital infidelity is more or less the same between men and women.[23] Interestingly, several studies argue that a woman is more likely to be unfaithful during her narrow fertile period around the time of ovulation.[24] Especially because ovulation is concealed in humans, this bears the mark of a deeply ingrained subconscious instinct: "I've got a good mate, but now's my chance to get some more diversity in my clutch—and maybe even some better genes." Remember, the evolutionary drive is not just to have children, but to have successful children.

The concealed nature of human female ovulation is peculiar indeed. In the other apes, the fertile period of a female is rather obvious. Their genitals become engorged and red. You can spot them a mile away. This is very useful for the husbands because they know when to mate with and guard their wives. As mentioned earlier, in baboons, mate guarding only takes place in the fertile period. After that, the males do not bother to guard their mates.

In contrast, human females do not advertise their ovulation at all. It is totally obscured, even to themselves. Why? This seems imprecise and inefficient. Without knowing when she is ovulating, a male would have to mate with her throughout the year and guard her constantly . . . bingo! This is the leading hypothesis of why ovulation is completely cloaked in humans.[25] The uncertainty about the timing of ovulation keeps the men interested in the females all the time. A male needs to stay close by to keep other males away, and he needs to mate with the female frequently to increase the chance of hitting the reproductive jackpot. This frequent mating strengthens the pair bond. Could concealed ovulation have been crucial to the genesis of the human family as we know it?

Not so fast. Concealed ovulation also means that adulterous suitors will be constantly interested as well. Since they do not know exactly when she is fertile, any quick dalliance is worth a try, and remember, she has incentives to get a little on the side. She could be protecting her children from a potentially murderous future alpha, or she could be shopping around for better genes or more diversity among her children. Either way, by hiding her ovulation, she keeps them interested.

Why all of this talk about infidelity in a chapter about jealousy? Key to the modern experience of jealousy is that both men and women have the demon inside of them, driving them to seek extramarital sexual activity. There is little to be gained by pretending that is not true. Many have conquered that demon by the sheer will of their conviction. Others have chosen not to fight it at all and instead choose a bachelor(ette)'s life or an open marriage. The great majority of us, however, are stuck in the middle somewhere, doing our best to maintain our promises of fidelity in the face of temptation, while also suspiciously guarding our spouses from those same temptations. In other words, we are at war with ourselves and with each other and therefore doomed to live jealous lives.

We humans are utterly convinced that we have evolved the ability to make our own choices and control our own destiny. We can choose our mates, our lifestyle, our commitments, and our fidelity (or lack thereof). Because of that, we like to think that our relationship-based emotions have become too sophisticated for an overly simplified look at jealousy. In humans, we tend to think that romantic jealousy must be much more than merely an attempt to protect reproductive investments.

Some data argue that we may be just as simple as our fellow primates. Researchers at Northern Illinois University conducted a study of many hundreds of college-age students.[26] Some were in relationships; some were not. Some were gay, lesbian, or bisexual; others were straight. Respondents were walked through several hypothetical scenarios and then asked to report how jealous they would feel in those situations. There were three hypothetical people: the respondent was the person participating in the survey, the partner was the intimate partner of the respondent (real or imagined), and the rival was the person tempting the partner into an illicit sexual affair. The survey involved scenarios in every possible direction: a single straight male whose imaginary girlfriend was being seduced by another man, a partnered gay man whose boyfriend was being seduced by a woman, a lesbian being seduced by another woman, and so on. You name it, they tested it, and respondents were asked to rate their jealousy and describe it as sexual versus emotional (romantic) in these various circumstances.

The surveys revealed that sexual jealousy spiked most when a woman was tempting a man or a man was tempting a woman, regardless of the sexual orientation of the threatened party. When it came to romantic jealousy (called emotional jealousy in this study), the results were more in line with what we would expect. The lesbians were threatened by the advances of a female interloper, the straight women by the intrusion of another straight woman, and so forth. The study also found what has previously been reported many times before: heterosexual females are more susceptible to romantic jealousy and less susceptible to sexual jealousy than heterosexual males (although this apparent phenomenon may have been erroneous all along, owing more to how women and men view the question differently rather than innate differences in how jealousy is activated.[27] Clearly, more research is needed.)

At first, I found this result odd. One would assume that a lesbian would always be most threatened by the attempted seduction of her wife by another woman. I would have suspected that flirtatious men would not bother her at all. When it comes to emotional jealousy, they do not, but the element of sexual jealousy is still there. Why? Perhaps it is because she knows that, technically, her wife and the intruding male could make a baby together. That possibility is unsettling to contemplate. Regardless of the emotional connection (or lack thereof) of an affair, the thought of our partner engaging in extramarital sex with a potential future co-parent boots up jealousy programming in our brain. Underneath the complexities of our emotional experience, reproductive potential is still an important aspect of human sexual jealousy, regardless of gender or sexual orientation.

But why should we have such great concern about reproductive fidelity in our mate? For males, it is rather simple. He does not want his mate(s) spending time, energy, and resources gestating, birthing, and caring for other males' children. Any infidelity by her is a direct and potent threat to his reproductive success, plain and simple, because it takes time and resources away from his children. He has a strong reason to try to stop it, and jealousy is the drive to do that.

For females, things are a little more nuanced. Why would she care about her man sleeping around? Once again, it comes down to his investment in the success of her children. This is where species differences in reproductive behavior come into play because some fathers, such as humans, invest a great deal in their children's success, and in other species, males do no parenting whatsoever. For species that do show paternal investment in child-rearing, the females know that if their mates achieve outside paternity, his protection, resources, and so on will then be split among her children and those of some other female. Even in the harem lifestyle, where a male has constant unhidden sex with multiple females, there is extensive tension and jealousy among the common mates of a single male. In gorillas, this is part of why the silverback is necessary to keep the peace among them. Interestingly, if the "sister wives" are actual biological sisters, as is usually the case in lions, jealousy is not a problem.[28] This makes sense because sisters share genes; helping your sister is almost like helping yourself, genetically speaking. If the wives are not sisters—well, remember Seri, the tigress and jealous wife who killed her husband with the wandering eye?

In sum, sexual and romantic jealousy, like so much else about our nature, is about reproduction. It is a drive, an instinct, to defend our reproductive success and prevent our efforts and resources going to the reproductive success of others. You might now ask, "Yes, but what about those who have chosen to forgo reproduction or who are beyond reproductive age? Those people still get jealous. Why might that be?"

Keep in mind that our drives and tendencies are largely subconscious. Behavioral drives are messy, imprecise, and immune to being neutralized by other facts on the ground. Although conception for procreation is not the only function of sex in many species, it is nevertheless a key focus of the sex drive. And yet, if you vasectomize a male, his sex drive is not affected in the least. Similarly, when a young married couple uses artificial birth control, the sex drive does not "know" and is as strong as ever. It is the same for jealousy. Just because a couple has chosen a child-free life does not mean they will be free of sexual and romantic jealousy. We cannot simply flip a switch and deactivate the jealousy program that is written into our mammal brains.

In sum, human sexual and romantic jealousy, while different in context, both harken back to mate guarding and defense of reproductive investments. As such, jealousy is one of our more basic emotions, whether we like it or not.

SOCIAL JEALOUSY IN ANIMALS

The biological forces underneath sexual jealousy seem pretty straight-forward. Males do not want to suffer reduced paternity or spend time and resources on other males' offspring. Females do not want the father of their children spreading his attention beyond the homestead. On the other hand, jealousy of other kinds of relationships—friendships, for example—seems wholly different. That kind of "social jealousy" is not found in the animal world, is it?

Of course it is. It is common knowledge that dogs get jealous very easily and very often. Any time a dog owner gives attention to another dog, there will be a jealous reaction. It may involve sulking, ignoring, or even depression, but more likely, the jealous dog will become aggressive toward the interloper. My own dog, Bruno, is a perfect example of this. When I would go into my former roommate's bedroom to pet her cats, Bruno would sit in the doorway and watch, clearly agitated. As I pet the cats, he would whimper and whine and periodically jump at the gate. If I took things further and uttered an enthusiastic, "Such a good boy!" to the cat, this was more than poor Bruno could take. He would go into hysterics, jumping and barking and howling. In short, he flew into a jealous rage.

Social jealousy has been observed in just about all companion animals. Whenever they perceive their owners giving attention to a possible replacement or rival, jealousy is usually the result. With sexual jealousy, animals and humans are responding to the very real danger of reduced reproductive productivity, but with seemingly innocuous social threats, why the obnoxious response?

To understand this, we must look at relationships in the animal world with an ice-cold eye. Whatever else animals feel, whatever attachments they form, and whatever grief they experience, they view their social relationships as commodities. For a given animal, his relationships have real value for his fitness—his survival and reproductive success. Animals make friendships and alliances that benefit them. Pair-bonding species select their mates carefully based on "what is in it" for them. Even the relationships between parents and children can be viewed in terms of value—the value added for survival and reproduction.

I know this is a bucket of cold water poured over our warm, fuzzy feeling toward animals, but animals really do approach their relationships as merchandise, as tools to help them succeed. Some species drop their mates when a better one becomes available. Friendships and social alliances are disrupted as the young and ambitious move their way up the ranks. Parents can abandon or even kill their own children when they become a liability in some way. For example, if a burdensome child is consuming resources that could go to healthier or more socially successful siblings, what is a parent to do out there in the harsh world?

With this in mind, social jealousy makes perfect sense. It serves to protect relationships that are of social or reproductive value. Take dogs, for example. Whether they view themselves as alpha or they are submissive to someone else, social rank is important. Your dog is jealous when you give affection to another dog because she is fearful of losing her place in the social structure. Each relationship has value in the dominance hierarchy, and rank matters to her. This is because higher-ranked wolves are more reproductively successful, dominate the food resources, and so on. It does not matter that your dog is spayed and has no shortage of food—the drive to jealously guard social status is deeply ingrained.

Accordingly, dogs view various relationships differently. It is not uncommon for dogs to favor some family members over others. Further, when strangers are introduced (human or animal), the same dog may react warmly to some but be hostile to others. It is not just unfamiliar individuals who evoke jealousy. Many a budding relationship has been quashed because the dog simply did not like the new girlfriend. Furthermore, in multiple-pet families, some dogs form close alliances, while others merely tolerate each other (on good days). Once again, these differences are usually best understood by viewing each relationship as a commodity. Dogs continually but subconsciously evaluate what each person or pet provides.

For example, if a dog recognizes a clear alpha in the house, she often attempts to maintain a close alliance with that person (or dog!), unless that individual rebuffs the attempt or repeatedly harms her. For better or worse, some breeds of dog respond to tough corporal punishment with staunch loyalty. This is not because they are masochistic. It is because they know that it is good to have powerful friends.

The one exception to this rule is the fact that many dogs are gentle and attentive with young human children. This would seem to be a conundrum because what advantage could it bring to befriend the weakest member of the pack? This is probably a quirk of dogs. Remember that their genetic behaviors are not solely those that were inherited from wolves. They are also the product of recent intense selection that honed certain work behaviors for the benefit of their human domesticators. Dogs have been bred with instincts and abilities to perform trained tasks, including herding, guarding, and protecting. This ingrained some special protective instincts toward little ones. (Also, I think it is pretty obvious that dogs know that kids are a ready source of dropped food. There is little to be gained from biting the hand that feeds you.)

SOCIAL JEALOUSY IN HUMANS

The value-threat interpretation of social jealousy translates well to all social species, including humans. We, like dogs, feel threatened when someone encroaches on a relationship that we value. For relationships that we do not value that much—eh, who cares if they went out with other friends and neglected to call us. Dogs do not show much jealousy if a member of the pack that they are not closely bonded with shows affection to another dog or human. If you scan the discussion in the last few pages about social jealousy in animals, you will probably not struggle to imagine the human parallels.

Most of us are not comfortable thinking of our various relationships as commodities. Because we hold our friends and family so dear, we refuse to accept that, even very deep down, we view them only in terms of what they provide for us, which is fine. There is an "out" that allows us to apply this reasoning to humans in a way that does not seem coldhearted: our jealousy reaction is tripped any time we experience or perceive encroachment on a valued relationship. Where that value comes from is totally up to us. Some of us value friends who are good listeners. Others are so-called fixers and prefer to surround themselves with people who need their help. Still others value friends that make them laugh and share common interests. It is not coldhearted to say that we value our friends. Their loyal friendship is important to us. Our friends and family cheer us when we succeed, help us when we struggle, laugh with us in good times, and provide a shoulder to cry on in bad times.

That is all fine and good, but like I said, we are done with the mushy stuff. Life on planet earth is not all sugar and spice. The harsh truth is, many humans behave very much like our animal relatives when it comes to their relationships with friends and family. All of us know people who seem to view others only in terms of what they can provide. Sometimes we call a person like this a user or two-faced. Some people are real jerks and seem to have no real loyalty to anyone or anything.

The maddening part is that not all of these shtunks end up in prison or penniless on the street. Some of them succeed in life. Some of them *really* succeed. They may suffer the occasional personal setback, but in the corporate dog-eat-dog world, being genuine and gentle counts for very little, and having high-placed friends counts a great deal. While this may seem like a product of our hyper-developed, artificially generated civilization, the sobering reality is that the corporate world is little more than the law of the jungle. It is just being applied in conference rooms and cubicles, rather than rain forests and savannahs. If anything, we are seeing a regression of sorts. Corporate piracy and vulture capitalism are not less in tune with our natural savage state; they are more so.

Forget about the jerks. Even among regular folks, everyone wants to be friends with famous or successful people. All things being equal, the more rich and powerful someone becomes, the more bombarded he or she will be with friendship solicitations. There are nonbiological reasons for that, of course, but I cannot help but think that our biology is at work as well. This is the "friends with the alpha" phenomenon. Sadly, the reverse is also true. Anyone who has been down on his or her luck will confirm that it often feels like many friends—and sometimes even family—have abandoned him or her. Ouch. I do not mean to depress everyone by claiming that we are all callous. I have always felt that understanding our biology (and thus our psychology) can actually help us rise above some of our more cruel instincts.

When it comes to social jealousy, humans behave pretty similarly to other animals. Although we do not usually snarl and bark, we might as well. (I actually know of a situation when rivalry between two postdoctoral fellows in a research lab deteriorated to the point where one actually hissed at the other. True story.) We are jealous in our friendships, interactions at work, and even our parental relationships. Even if social relationships do not really offer much value to our reproductive success anymore—although I would argue that they do—this intrinsic part of our nature remains.

Viewed through the biological lens, a work colleague receiving praise is clearly a threat to your social status because the professional world is often competitive. Some employers may try to pretend that it is not in order to promote cooperation, but we all know that it really is. When a colleague gets praise or a reward, we feel that this diminishes our standing with the alpha—the boss or supervisor—so we get jealous of her. Similarly, when a friend in your group is routinely the benefit of random good fortune, and you are not, it bothers you, right? This all stems from our competitive nature and our desire to maintain our position in the social ranking. This is social jealousy.

Ever heard of sibling rivalry? Same thing. Of course we all want the best for our brothers and sisters . . . as long as they do not outshine us. The embarrassing truth is that we can all be a little jealous of the attention our parents pay to our siblings. In my own family, there is a running debate about who is the "golden child" of the five of us, the one that can do no wrong in the eyes of our parents and receives the most love and support from them. (For the record, it is Darren.) As strange as it may sound, I would argue that this all comes back to resources and reproductive success. As animals, we instinctively recognize that our parents only have so much to give and any time and resources given to our siblings is time and resources not given to us. Cooperation and competition are always in tension and never more so than inside one's own family.

It is also delightfully sensible that sibling rivalry almost instantly gives way when an outside threat is perceived. It is a common joke that our older siblings can pick on us incessantly, but if someone else dares to, watch out! That is certainly true for me and my older siblings. It is a perfect metaphor for the tension between competition and mutual cooperation that happens in nature. We have powerful antisocial instincts to succeed at all costs, reserving as many resources for ourselves and our offspring as possible (see chapter 8, on greed), but these are balanced by the prosocial instincts to cooperate. Within a family, this is magnified. Intrafamily competition is normal, and social jealousy rears its ugly head in the form of sibling rivalry. However, within the larger herd, the forces of kin selection take over, and siblings see each other as a combined unit in competition with the others. This is as true for kangaroos as it is for humans. We are all jealous beasts.

FURTHER READING

Note: These two works take very different views on the origin and nature of human jealousy and are presented here as opposing theories.

Buss, David M. *The Dangerous Passion: Why Jealousy Is as Necessary as Love and Sex.* New York: Simon and Schuster, 2000.

Campbell, Anne. *A Mind of Her Own: The Evolutionary Psychology of Women.* Oxford: Oxford University Press, 2013.

8

DARKER STILL

Envy, Greed, and Power

THIS CHAPTER CONTINUES our disturbing journey into our dark side. Envy, greed, and power are linked because they all revolve around the acquisition of resources. Resources can include money, possessions, positions, or people—whatever serves our needs and desires. We all have a desire to obtain these things, and sometimes we go to great extremes to get them—even if it means violating our own morals.

Envy is defined as coveting specific resources that others have. Herein, I shall focus mostly on possessions, wealth, and other material things, not relationships. (We covered relationships as resources in the last chapter.) Greed is the relentless pursuit of resources, above and beyond basic necessities for survival and comfort. Power is control over resources, including and especially people. Dictionaries have different word choices, but those are the definitions that frame this chapter.

We are all guilty of envy, greed, and the pursuit of power at certain times and in certain circumstances. We all want to do well for ourselves, and we all want the independence to make choices without economic or social restrictions. We all want a big home, a nice car, toys and gadgets to entertain us, a generous salary, plenty of vacation time, a beautiful husband or wife, darling and obedient children, supportive friends, and so forth. We all want those things, and that is not a bad thing, per se. Envy and greed start out as normal ambitions; there is nothing wrong with seeking and getting the basics in life.

Envy and greed become evil or sinful when the pursuit of resources becomes unhealthy, excessive, and insatiable. We begin to harm others, either intentionally or unintentionally, as we put our own desire for more stuff ahead of the needs and desires of others, including those who already have less than we do. Particularly with greed, we are talking about a callous and cruel disregard for the needs of others.

Delineating between normal ambition and ruthless power-mongering is a bit arbitrary. There is no clear line in the sand, only shades of gray. I think that we can all agree that working hard at your first job out of high school and then asking for a reasonable raise is an example of totally normal and healthy ambition. Similarly, nearly all of us would label a billionaire hedge-fund manager as greedy and evil if he cut benefits from his employees just to save a few bucks for himself. At the two ends of the spectrum, the moral judgments are more or less unanimous. Cases in the middle, however, are harder to label, and reasonable minds can disagree.

Greed is also a double-edged sword in nature. Healthy competition for resources is good for the long-term survival of a species, just like economic competition is good for the economy and distribution of wealth. Similarly, greedy and power-obsessed animals can threaten the survival of a species just like unbridled greed can do great harm to an economy and its political structures. Greed factors very prominently in the same dance of competition versus cooperation that has been discussed in other chapters. Greed pits the personal ambition of individuals against the success of others.

IT IS ALL ABOUT RESOURCES (AND SEX)

As mentioned in the previous chapter, in social animals resources bring status, and status brings reproductive success. We have seen how relationships, whether social or sexual, are interconnected with status. Possessions, too, can enhance social status.

First, a familiar story: Remember the cliff swallows from chapter 5? These birds live in very large colonies with an intricate social structure. One aspect of that social structure is the building of nests. Their nests are, shall we say, unique. They are made of a muddy paste that is allowed to dry in a tube shape and stuck onto the cliff face, sometimes even underneath a ledge. Picture a very large hornets' nest-like tube, big enough for a small family of birds, stuck to the side of a stone cliff. The paste is made from mud, sand, sticks, feathers—whatever they can find.

Given the herculean task of building these nests, it is not too surprising that the size, quality, and location of a nest bring social status with it. For example, nests nearest to the watering hole, lake, or ocean are the most coveted. This is not just because of the easy access to drinking water but also because the water is where much of the social interaction takes place. If you remember, although the swallows form stable mated pairs, there is a lot of extramarital courtship and sexual activity as well. Much of that takes place around the watering hole or shoreline. It is a prime spot. The further you get from the water, the lower your status. It is just like the real-estate market in any major city: location, location, location. Because status equals sexual desirability, there is a pretty strong correlation between proximity to the watering hole and reproductive success in swallows.[1]

This is where envy comes in. Those with the nice nests in the fashionable part of the colony are sometimes the targets of nest theft. A mated pair of cliff swallows may envy another pair's nest so much that they actually work together to forcibly evict them from their home and steal it.[2] Of course, such a heist must be executed carefully. The operation must take place when only one member of the pair is home defending it. (Swallows are not so foolish as to leave a nest totally unguarded for very long.) Further, there is no point trying to steal from a larger pair of birds, since they will simply take it back. The thieves could lose their own unguarded nest in the meantime.

The point here is that envy, the desire for someone else's things, is clearly present in this species, and with good reason. Having a better nest actually gets you something: more offspring because of your social rank. For both the males and the females, having a prime nest means that they will be more attractive to potential copulation partners at the watering hole. For the males, that means more offspring. For the females, that means baby daddies of higher status. Both cases lead to enhanced reproductive success. Natural selection would thus tend to reward both envy and home theft. Whoever said that thieves never prosper was not an ornithologist.

However, there is a downside to having a highly desirable nest. Besides the risk of home invasion, prime nests are also frequently the "victims" of brood parasitism—the placement of others' eggs in their nests in order to "mooch" their parenting efforts. Sneaky cliff swallows have evolved a keen sense of figuring out which nests are the best for rearing fledglings. It is more than just the sexiness of the residents and the location of the nest. Some nests are infested with egg parasites that damage or destroy the young eggs before they can develop and hatch. Somehow, cliff swallows can detect the level of infection in other nests. They use this knowledge to help them decide where to place their eggs if and when they decide to parasitize the parental efforts of others. Nests that have lower levels of egg parasites will suffer higher rates of brood parasites.[3]

You may recall that I railed against the very idea of brood parasitism back in chapter 5 and instead argued that cliff swallows simply employ a system of communal parenting. I am not contradicting that claim here. It must always be remembered that competition is still present, even in cooperative social systems. In cliff swallow colonies, extra-pair copulation and egg transfer are part of the social fabric, but that does not mean that they are always harmonious. There is still competition, resistance, and struggles. It is just like dominance struggles among rams or dating in humans: there are often casualties.

There is no "pure" cooperation in the animal world. It is always a balance. Even though brood parasitism is common among cliff swallows, it does not mean that it is all perfectly egalitarian and selfless. No bird (or human!) likes to carry the weight of others, even though he may accept that he has to do it from time to time in order to live in harmony. While I emphasized the cooperative side in chapter 5, I admit now that plenty of sneakiness and selfishness is involved.

This sounds awfully human to me. Each of us is a swirling mass of good and evil drives. We care about others, especially those we love but also strangers in need; yet we also work to get ahead of others. Most of us see nothing inconsistent about volunteering at a soup kitchen on Sunday and working tirelessly to undercut and defeat a competing company on Monday. Both the soup kitchen and the competing company are filled with our fellow humans, are they? Similarly, the difference between a teammate and a competitor is often found by analyzing whether or not we gain from their success. If helping John Smith helps me, we are teammates. However, a teammate can instantly become a competitor when a promotion becomes available and only one of us can get it.

Envy and greed, like jealousy, is the shifting of our focus from the cooperative to the competitive. Humans, like all social animals, have long known that working together promotes our common prosperity and makes available more resources for each of us. Within that scheme, however, there is nothing guaranteeing that those resources will be partitioned equally. Envy and greed motivate us to acquire and consume more than our fair share and, in so doing, ensure our genetic legacy through reproduction.

RICH GUYS GET ALL THE GIRLS (AND VICE VERSA)

Although we may not be aware of it, one key aspect of our motivation to get an education or a good job is to make ourselves desirable to a spouse or partner. Think about it. When meeting potential dates, "What do you do?" is one of the first questions asked—if not the first. "What does he do?" is also the first question that parents ask their children when they learn they are dating someone new. Why would that be? Surely not because everyone thinks that accountants are better parents than scientists or some other nonsense. The blunt truth is that we ask the question, "What do you do?" because we want to know about the earning potential of the suitor. Truth be told, we would probably rather ask, "How much do you have?" but that would be a bit too obvious.

There is a cold truth behind the reality that everyone wants their children to bring home doctors. Face it: if we wanted our children to bring home people that are more likely to be caring and dedicated parents, there would be much more cachet in saying, "My daughter is dating an elementary school teacher!" Admittedly, financial resources matter more to some than others, in part because some people fancy themselves as the breadwinners and so will concern themselves with other traits when searching for a mate. Still, I do not think it is a stretch to say that having a lucrative profession is seen as highly desirable in the search for mates.

Of course, mate selection is an aspect of human society in which gender roles and social conditioning still have a hold on many. Nevertheless, the general connection between attractiveness and wealth is undeniable.[4] Bill Gates has a large number of female admirers (males, too) who express an overwhelming physical attraction to him. Yes, Bill Gates. Of Microsoft fame. What do you think they are so attracted to? It is difficult to imagine that he would get a second look from those people if it were not for the fact that everyone knows he has an enormous stock portfolio. Donald Trump never struggles to score beautiful women, and we know it cannot be because of his looks or his personality. Rich and powerful people are seen as more attractive than they would be otherwise. We all know it.

Interestingly, there may be some symmetry in the relationship between wealth and attractiveness. It has been suggested more than once that base level of attractiveness can also impact future success. Jerry Seinfeld once joked, "You don't see many handsome homeless." Callous jokes aside, there may actually be some truth to this premise. Attractiveness has been shown to have an impact in simulated personnel actions such as hiring and promotion.[5] Attractiveness and self-perceived attractiveness can even impact everyday exchanges in ways that benefit people who are good-looking or at least believe that they are.[6] Sadly, this phenomenon likely begins well before adulthood, and attractiveness can have bearing on academic success of adolescents in certain instances.[7] Teasing out the relative contributions and cause-and-effect relationships of attractiveness, self-confidence, and the expectations of others is extremely difficult, but suffice it to say that there is a well-documented link between attractiveness and wealth/power, and this link likely operates in both directions.

In this light, the evolutionary origin of greed and lust for power is rather obvious. The more wealth we acquire, the more attractive we will appear. The more attractive we are, the more attractive (and numerous) will be the mates that we attract. Therefore, wealth and power have the potential to bring us not only more children, but better children, at least in terms of attractiveness, if we assume that attractiveness has some genetic basis. These more and "better" children will help propagate not only the genes for attractiveness but also the genes for greed and appetite for power. It is a rather simple and self-reinforcing cycle. Of course, this drive would be held in tension by the advantages offered by cooperation, but the advantages enjoyed by the greedy are obvious. Just take a look around the world today. It is no wonder that humans came up with the concept of karma.

What about animals? Do wealth and resources actually lead to more and higher-quality sexual partners? Not surprisingly, the answer turns out to be yes. In many species, one sex does the "courting" and the other does the "choosing." There are many ways that an animal can court another. Some sing, some croak, some display colorful body decorations, some fight, some dance, and so on. However, some try to impress the opposite sex through their acquisition of resources. It is not the same as "purchasing" sexual access, although animals do that, too, as we saw in chapter 4. Some animals build an empire of sorts in order to attract a choice mate, whatever that entails for that particular species. Some examples will illuminate.

Many birds employ the phenomenon of territorial display. Most hummingbird species are a good example of this.[8] In the spring, male hummingbirds will stake out a territory and then make mating calls to attract a female to roost. However, the females are not attracted to the males based on the quality or volume of the song, as are some other birds. The calls just grab the females' attention. Once a female hears the male, she comes over to him and checks him out. It is more like a catcall than a courtship song. However, once she starts looking, the things that make a male attractive have little to do with what he is and everything to do with what he has. The females are most interested in a male with a large territory with lots of food sources and places for nest-building—in other words, a wooded area with diverse vegetation.

Under this system, the males have every reason for ruthless greed. In North America, most hummingbird species are migratory, and the males arrive back in the north two to three weeks before the females. This gives them time to establish their territories. Each male tries to stake out as large and rich a territory as possible, with lots of flowers and trees. However, the larger and more densely vegetated the territory is, the more difficult it is to protect from neighboring males looking to expand their territory. A male will protect his territory by standing guard on a high branch. Intruders are then warned with distinctive calls and wing-beating sounds. If the "verbal" warnings are not enough, the territorial male attacks the trespasser. To do so, he floats up to a very high altitude and executes a dramatic (and dangerous!) nosedive attack. The intruder is rarely injured and usually just moves on.

Most of the male-male fighting takes place before the females arrive. By the time they do, the territories are established. A female spends just a few days visiting various males, and she makes her choice almost exclusively based on the size and quality of his territory.[9] Interestingly, there is a debate about why the territory matters to her so much. Some argue that the territory itself is an attractive place for her to live, build a nest, and raise her chicks. However, since most hummingbird males stop guarding the territory after they have mated and contribute nothing to the nest, eggs, or hatchlings, some see the territory-based selection process as an expression of sexual selection. Since large and rich territories equal reproductive success, females choose baby daddies that are good at defending them in the hopes that their sons will also have that talent. It is a version of the "sexy son" theory that females choose desirable mates at least partially because they want desirable sons to promote their own reproductive success. In this case, however, desirability comes from your real estate, not your looks.

A slightly different twist on this hypothesis claims that the ability to stake out and defend a large and desirable territory is a good proxy measurement of the overall health and vitality of the male. For example, a bird needs good eyesight and visual processing, as well as a keen sense of smell and knowledge of trees and flowering plants, to assess the suitability of a patch of land. He needs certain cognitive functioning to make judgment calls about how much territory to try to defend: too large and he cannot defend it; too small and he may not impress the ladies. He must be alert with good vision and hearing in order to detect intruders before they have established a strong position in the territory. Finally, to pull off the impressive nosedive attack, he must be strong and fast, with good muscle coordination and integration of senses such as vision, tactile perception (to sense wind, air resistance, and contact with leaves), and the sensation of position, equilibrium, and balance. All in all, I would say the selection criterion used by female hummingbirds—the size of a male's territory—is a pretty reliable summary measurement of the overall quality of the males.

It turns out that hummingbird-style territorialism is common, though not universal, among birds.[10] Woodpeckers, mockingbirds, cardinals, and most other "backyard birds" are territorial, with the males usually partaking in the defending and the calling. Interestingly, in most of these birds, the males stick around after insemination and actually contribute the parenting equally or nearly so. (Hummingbirds are in the tiny minority, among birds, for having deadbeat fathers.) Accordingly, the territory is even more meaningful to other birds than it is to hummingbirds. The males actually utilize much of their territory, together with their mates, foraging for nest materials and food for the family. That the desire for prime real estate is retained in hummingbirds even though it is not really needed indicates that this is a deeply ingrained instinct in birds.

Red-winged blackbirds are even greedier than hummingbirds. In this species, the territorial males also attempt to stake out territories, but square footage is not as important as the richness in fruit and sources of water, which are necessary to support the brood of hatchlings that the males hope to sire there. However, you would think that a male would be satisfied with a lush territory that provides ample fruit and water to support the largest possible nest. Nope. A male will continue to seek, fight for, and defend more and more territory. If he is successful at that, he may attract a second female, who will establish another nest. After that, he will go on searching and fighting. He is never quite satisfied. In this scheme, the most hapless male blackbirds end up with zilch, and the most aggressive and successful will father as many as ten nests with three to four eggs each. Most of the males end up with one or two mates, making this species a rare exception to the monogamy exhibited by most birds.[11]

The most interesting thing here is that the polygyny of red-winged blackbirds is directly related to resource acquisition. There is a near-perfect correlation between how rich a male can make himself and how many wives he will get. This is a prime example of the connection between greed and reproduction.

Using acquired resources to attract mates is not unique to birds. We already talked about how most male fish build nests to entice females to deposit their eggs.[12] The females are most attracted to males that have built (and have been able to defend) large and elaborate nests. In many species of fish, there is no "married life," and the females do no parenting. They simply squirt their eggs, and that is it. Judging a male by his nest works well because, presumably, the ability to build and defend a nest is a measure of the health, strength, and skill of the male. Health, strength, and skill are things that all fish mothers want their children to have. Also, in theory, defending a nest prior to spawning is not all that different than protecting the eggs afterward. For fish, judging a male by the home he has built makes a lot of sense.

OTHER TYPES OF ANIMAL GREED

Although most examples of animal greed involve the acquisition of resources for the direct purpose of reproduction, there are a few additional reasons for animal greed. Take, for instance, the wolverine. This aggressive and powerful member of the weasel family has a reputation for being quite the bully. Wolverines can successfully fight off and chase away much larger animals, including bears, cougars, and moose (!). There are even stories of a lone wolverine fending off an entire pack of wolves, each one of which probably having outweighed the wolverine. They also have a penchant for being smelly. Sometimes called the skunk bear, wolverines frequently spray their territory, their food, their mates, their offspring, and even themselves with a mist from their anal glands. Given their toughness, other animals are well advised to steer clear of the musky scent of the wolverine.

Not surprisingly, wolverines are expert hunters, rarely preyed upon, and comfortably at the top of their food web. Because many other predators seek the same food sources, wolverines have become fierce competitors. They are known to chase other scavengers away from a carcass, and they have no shame in stealing a hard-earned kill from a smaller wolverine or even a different animal. They are voracious eaters, which gives rise to their various names in other languages such as "glutton" (in French), "gluttonous badger" (in Romanian), and "fat belly" (in Finnish). In fact, the scientific name of the wolverine is *Gulo gulo*, from the Latin word for gluttony.

Although wolverines sound rather like playground bullies, this is all pretty standard food competition. Where does the greed come in? Well, after a wolverine has eaten all he can, whether from his own kill or find or from something he has stolen from some unfortunate schlimazel, he will actually spray the leftover food with his marking scent. Biologists once thought that the wolverines were marking the food to protect their next meal of leftovers. However, this does not seem to be the case. Wolverines rarely return to their leftovers. Sure, the distinctive wolverine scent alone is probably enough to dissuade many animals, but it turns out that the spray of wolverines, unlike that of skunks, for example, is highly acidic. By spraying noxious carboxylic acids onto the leftover food, the wolverines actually accelerate the spoiling process.[13] To summarize, the wolverines consume all they can fit into their stomachs, and then they try to spoil any leftovers so that other predators and scavengers cannot eat them. This fits part of our description of greed. It is not just about acquiring things; it is about having more than others have.

Those not comfortable assigning the term greed to these small weasels may counter that this is just a good competitive strategy. If an animal is in constant competition with other animals for the same food sources, there is an advantage in not feeding the competition. By leaving leftovers behind, the wolverines would be helping future competitors live to fight another day.

My response to that is, "Bingo!" Greed is precisely that. It is an intense competitive strategy that goes beyond just getting what we need. It is also about preventing others from getting what they need. Greed makes us see everyone else as competition and drives us to measure what we have against what others have.

What about primates, our closest relatives? The many species of primates exhibit widely different social structures. On one hand, bonobos have a highly stratified social structure with a strict rank order that determines everything from mate choice to partitioning of resources. On the other hand, there are orangutans, which are mostly solitary as adults. Gibbons tend to form nuclear families, while gorillas live in harems. When it comes to sharing of food and resources, primates pretty much run the whole spectrum.

Researchers at New York University discovered that rhesus monkeys display similar behaviors to humans when it comes to risk tolerance and wealth.[14] These scientists used water as the proxy for wealth, as water resources are important to rhesus monkeys, which are adapted to arid habitats. In this model, greed—the "thirst for money"—is represented by actual thirst. Over many days, the monkeys were trained to make choices that could lead to them receiving water. There was a "safe" choice that gave them a small amount of water no matter what, and there was a "risky" choice, which gave them a 50 percent chance of getting a large amount of water and a 50 percent chance of getting nothing.

After the monkeys had been well trained on the gambling system, the researchers presented them with the choices when they were either well hydrated or a bit dehydrated after a short period of water deprivation. This was to simulate wealth and poverty. Researchers measured the blood osmolarity (a measure of hydration) of the monkeys and then provided the monkeys with the water choices. Interestingly, the thirstier monkeys nearly always opted for the safe choice: a guaranteed small amount of water. The well-hydrated monkeys tended to take the fifty-fifty choice. With plenty of water already, they were willing to tolerate the risk of getting nothing for the chance of getting a big payoff.

While this study was, first and foremost, a proof of concept that rhesus monkeys can be trained for experiments involving game theory and wealth apportionment, it revealed that primate and human psychology share features when it comes to risk and wealth. The less you have, the more risk averse you are. The more you have, the more you are willing to risk in order to get even more. Economists have known this about humans for some time.[15] Those who are already wealthy, or whose proximate family is, are more likely to take risks such as starting a new business, changing careers, or putting large sums of money into venture capital. It is not just that they have more to risk; they are looking for the big payoff. For those of more humble means, this seems greedy. If you already have a decent living, why not go for safe choices that will ensure you retain your comfortable lifestyle?

Wall Street is the perfect example of this. Wealthy investors start hedge funds or engage in rampant speculation about the future of specific markets and commodities. As we have observed over the past decade, such wild speculation is incredibly risky. If left to their own devices, an elite group of greedy investors can take down not just themselves but the entire global economy as well. The most aggravating part is that they were wealthy already!

While it may be oversimplifying to compare human greed with the choices that thirsty rhesus monkeys make, there are definitely parallels, which speaks to an underlying common thread: wealth invites risk; poverty leads to risk aversion. That behavioral feature is likely present in all primates and could have evolved into human greed in our lineage.

THE EVOLUTION OF GREED
AS A BEHAVIORAL PROGRAM

How did greed evolve? Consider the examples of animal greed that I have discussed. The example of territoriality in hummingbirds shows that, for some animals, the aggressive pursuit and dominance of terrestrial resources has a direct benefit for them. They have every reason to be greedy. Further, not unlike human greed, that appetite for land resources is purely selfish. The birds do not actually need that territory, and they do not end up using it for anything. They want it simply because they do not want others to have it.

In fact, if each male hummingbird would be more content with a more humble plot of land, the environment might even be able to support more hummingbird nests. The individual pursuit of success seems to limit the potential of the species. This is not an uncommon outcome because evolution is not goal oriented. Negative features are sometimes perpetuated by their own popularity. Many species have evolved themselves into evolutionary dead ends. After all, 99.9 percent of all species that have ever existed on the planet are now extinct. Fierce and counterproductive intraspecies competition is no doubt a big part of that. Take heart: there is a reason that mammals now dominate the planet, with birds not far behind: these two classes of animals have made the greatest strides in augmenting competition with cooperation.

Is human greed like hummingbird greed? Is it largely about showing off, increasing our social standing, and impressing potential sexual mates? I would argue that it is, but for now, just say that a "greed program" does exist in the animal brain, and it is reasonable to think that humans inherited this ancient program from our ancestors. The greed program is a behavioral suite of observations, calculations, and desires associated with the acquisition and dominance of resources.

Greed in humans is not necessarily identical to that of other species because a behavioral program can be shaped by evolution over time, just like an anatomical feature. Consider the digits of the forelimbs in mammals. In different habitats and lifestyles, natural selection has shaped these digits into fingers, paws, hooves, fins, and wings in humans, dogs, horses, dolphins, and bats, respectively. The same anatomical digits are the common underlying chassis of those wildly different structures. Similarly, behavioral programs are like templates that can be modified in different lineages based on the adaptive pressures experienced in a specific time and place. In different kinds of animals, the greed program can be suppressed, enhanced, and tweaked as appropriate.

For example, it should be noted that not all bird species engage in territorialism like hummingbirds do. Those that engage in communal nesting, like herons, swallows, and most waterfowl, do not show any territorialism toward each other. They do defend their nests, but that is something else altogether. They do not even build those nests until after they have formed a mated pair. For most swallows, courtship involves song, not territory.

This does not mean that these communal birds do not have greed or that greed is not about reproductive potential for them. Remember the cliff swallows? They are greedy, but in a different way. They do not stake out huge territories in order to show off, but they will engage in nest thievery. They may try to pass their eggs to others so that they can lay more. And that is greedy. The greed program can manifest in a variety of ways, from territoriality to hedge funds and brood parasitism to corporate malfeasance.

Ergo, to dissect the programming of human greed, I turn back to experiments in our closest relatives, the primates. In possibly the most sophisticated look at how nonhuman primates behave in regard to wealth and resources, Professor Laurie Santos at Yale University has spent years teaching capuchin monkeys the concept of money using silver-colored tokens.[16] The monkeys use the tokens to purchase various items from the researchers, mostly their favorite foods. To help simulate a marketplace, they created a special enclosure where all transactions take place and certain research assistants become "salesmen" for specific items. By giving the monkeys a limited number of tokens and many options, the scientists can explore the decisions that the monkeys make and probe the effects of comparative pricing, salesmanship, and sales and discounts as they force the monkeys to make difficult decisions that reflect the value that they place on certain items.

Within this simple economic system—"monkeynomics," as it came to be called—some fascinating phenomena were observed. Not surprisingly, theft was one of the first things that the researchers noticed. Of course animals will steal food, toys, and tools from each other; that is nothing new. However, this was the first time it had been shown that animals would knowingly and intentionally steal currency in order to spend it on other things.

In other matters, the monkeys do all the things that humans do with their money. They consider the price of things before they purchase, they take advantage of discounts, and, when offered various choices, they budget their purchases to get as much out of their money as they can. As far as we can tell, they do grasp the concept of money and behave in ways identical to how humans behave in a similar market scenario.[17] (This was the same research group that documented the exchange of tokens for sex, discussed in chapter 4.)

The most interesting observation is that monkeys behave similarly to humans when it comes to risk and loss. This is an elaborate experimental scenario, so buckle up and bear with me. There is a bizarre feature in humans that has been known for some time: we are more able to tolerate risk in order to gain something than we are when we might lose something, even if the end results are the same. Here is an example similar to one often used by Professor Santos: suppose I give you twenty dollars. It is yours to keep if you want it. However, say I then give you a choice. You can do nothing and I will give you five dollars more. Or you can flip a coin and you get ten dollars if it comes up tails and nothing if it comes up heads. In this scenario, most people actually choose to play it safe, take the extra five dollars, and walk away with twenty-five dollars. It is something for nothing, so why risk throwing it away?

In another scenario, I give you thirty dollars. Then I force you to choose between two options, both involving the possibility of loss instead of gain. You can flip a coin: if it comes up tails, you lose ten dollars, and if it comes up heads, you get to keep the full thirty dollars. Or, you can play it safe and do nothing, and you have to give back only five dollars. Which do you choose? If you are like most people that have been placed in this scenario, you will choose the riskier option and opt to flip the coin. This is bizarre because the choices in the two scenarios are actually the same. In both cases, you are choosing between twenty-five dollars and a fifty-fifty chance of twenty dollars or thirty dollars. And yet, most people play it safe in the first scenario but take the risk in the second scenario.

This contradiction is a sign of inconsistent and irrational thinking. It is a perfect example of how our intuition toward making financial decisions is frequently flawed. We hold on to a losing stock, even though all forecasts tell us to sell. We are easily fooled by sticker prices that are heavily inflated only to be marked down as "50 percent off." We will not sell our house at a loss, even as taxes, bills, and cost-of-living expenses pile up. We do not measure the value of things solely on their own terms. We measure the value of things relative to what we spent on them or what they were supposedly worth previously. Humans are prone to behave irrationally when faced with losing money.

Professor Santos's research group arranged choices equivalent to those mentioned in the previous paragraphs. Using the monkeynomics system, they found that the capuchin monkeys exhibit the same risk aversion that humans do when they are faced with losing what they have already purchased.[18] This discovery tells us that the irrational way we behave when losing money is not human nature; it is primate nature. Capuchins diverged from our evolutionary lineage about forty million years ago, meaning this psychological quirk is an ancient and hardwired feature of the primate brain.

Furthermore, I would argue that this speaks to the evolution of greed in humans. Specifically, greed is about relative wealth. Greed means wanting your yacht to be ten feet longer than your neighbor's. Envy is being upset when your neighbor gets an even bigger yacht in return. It is a never-ending cycle. "Keeping up with the Joneses," it is sometimes called. We all do this, but my point here is that the reason may be because we have evolved to judge amounts of wealth and resources in purely relative terms.

GLUTTONY

Gluttony is the overindulgence and overconsumption of food and drink. It is similar to greed, in a sense, but focused specifically on foodstuffs and not in any way directed at others. Gluttony is overeating for its own sake, not to get ahead of others. It is a private and personal sin and something we are not usually eager to show off. In that sense, it is probably closer to masturbation than to greed.

Gluttony is something pretty much all of us are guilty of from time to time. We all love to eat. We especially crave rich, calorie-packed foods. Knowing that we "should not" often makes no difference at all. How often has your willpower collapsed at the sight of a scrumptious, moist chocolate cake? Our drive to eat is much more than just basic hunger. I think I speak for pretty much everyone when I say that if there were a magic pill that would allow me to eat anything I wanted in unlimited quantities and not suffer any weight gain or health risks, I would eat about six gigantic meals a day. And pecan pie. Lots of pecan pie.

There is a drive within all of us to eat, eat, and eat some more. Unfortunately, this drive also tends to focus on foods that are horrible for us. When was the last time you had an intense, mouth-watering craving for brussels sprouts? At our most basic level, we are built to crave high-fat, high-sugar, and high-protein foods. Sure, many of us grow into an appetite for more healthy foods and learn to shun the empty calories of milkshakes and soda, but if you believe that is the result of anything other than conditioning and training, you are kidding yourself. If children exhibit a more basic form of human emotions, their appetite for sweets and aversion to vegetables speaks for itself.

Gluttony is a feature we share with pretty much all animals. Anyone with dogs or cats knows how they can gorge themselves on treats, meat, and other rich and savory foods. Often, we have to carefully regulate the diets of our companion animals or else they will become overweight very quickly. The same is true for laboratory animals such as rats, mice, rabbits, fish, monkeys—you name it. Zoo animals, too. Great care has to be taken to select their diet—not just to include the diversity of food that they need to be healthy but also to regulate their intake so that they do not become morbidly obese. In sum, almost all animals, including and especially humans, if left to their own devices, will overeat to the point of extreme obesity. Why on earth is that?

Contrast this with the seemingly contradictory fact that you almost never encounter obese animals in the wild. Animals in their "natural state," that is, the environment they are adapted to through thousands of years of evolution, are most often trim or even skinny. When we put them in an artificial habitat, they will immediately balloon up if we are not careful. Why would this be? Could it be that the artificial nature of the simulated environment just is not right for them? Indeed, it was previously thought that the stress of captivity caused hormonal changes and nervous overeating. It turns out that that does not seem to be the main issue. You might also guess that the lack of proper physical activity and exercise is the culprit. Nope. Plenty of experiments and anecdotal experience have disproven both of those hypotheses.[19] So why do animals stay skinny in the wild but become obese in captivity?

The answer is a little disturbing. It turns out that animals in the wild are probably living in a near-constant state of intense hunger. As noted in the introduction, life on Earth is a pretty difficult experience for most animals. Life has been bustling on our planet for at least 3.5 billion years, and the animal kingdom emerged at least six hundred fifty million years ago. That is ten times longer than the amount of time that has passed since the last dinosaur died. During all this time, the proliferation of animals has allowed them to fill virtually every possible niche, in which they experience intense competition with each other. The great majority of animals in the wild live their lives teetering on a knife's edge between survival and death. There simply is not enough food to go around. The fact that all species tend to produce far more offspring than can possibly survive was one of the first key realizations by Charles Darwin, leading him toward his discovery of natural selection.

What does this have to do with gluttony? Because animals are locked in a vicious struggle for survival, they are wired for intense hunger; they seek food essentially all the time and will consume every last bit that they can. After all, who knows how long it will be before they get another chance? Only by gorging on food when it does become available do animals get the best chance of surviving to the next meal. A lion will eat fifteen pounds of meat in a single sitting; a snake will eat a meal that can nearly equal its own body weight. President Teddy Roosevelt told a tale of a school of Amazonian piranhas devouring an entire cow in minutes.

Humans also have the drive to eat at every opportunity. As soon as our stomachs empty, we want to eat again. Fruits and vegetables will do if nothing else is available, but we really want the good stuff: high-fat, high-calorie foods. Those are the foods that are most essential for surviving for days and weeks when food may not be available. The problem is that nowadays, food is available to us constantly. Our brains are not built for that. Sensible eating decisions and disciplined restraint are not part of our inborn psychological toolkit. Nowhere in the animal world is self-denial important for survival. Today's era of ready access to rich foods is a new experience for Homo sapiens, and there has not been near enough time to expect a change in our biology because of it.

It should be noted that it was previously thought that our relatively sedentary twentieth-century lifestyle was mostly to blame for the recent increase in obesity seen in Western populations. The idea was that, in previous generations, much more of the population earned their living through physical toil, and, prior to electronics, most recreation was physical. While these two phenomena do probably play some role, the idea that decreased physical activity is chiefly responsible for the recent obesity epidemic is now falling out of favor.[20] The availability of rich foods and our resulting calorie-rich diet seem to be the main culprits.

For the masses, it is only been a few hundred years that rich food has been this readily available, and that is only in the developed Western world. Before the Industrial Revolution, only an elite few could eat rich foods every day, and the rest were not much better off than animals in the wild. Indeed, being stout or plump was a sign of aristocracy and privilege until the early twentieth century.[21] We are now surrounded and bombarded with high-calorie foods, which goes against millions of years of evolution that have trained us to overeat whenever possible. Overeating was a great strategy when it was only rarely possible anyway, but now we are able to do it every day—multiple times. For most of us, our feeble willpower is simply no match for our physiology. As far as our bodies are concerned, at every meal, we are pounding on the energy storage, as if for a long winter when we may barely eat at all.

It is even worse than that. In addition to a tendency to eat, eat, eat and choose energy-rich foods, our bodies are also built to adjust our metabolism and fat-deposition patterns to easily gain weight and have difficulty losing weight. Anyone who has struggled with weight loss can tell you this. Weeks and weeks of dieting and exercise results in negligible weight loss, while a weekend eating binge can pack on a few pounds just like that. This is not an illusion caused by our cynical human tendency to see the glass half empty. Our bodies really do adjust our baseline metabolic rate in order to prevent weight loss and promote weight gain.[22] This is accomplished by ramping down our "extra" uncoupled thermogenic energy expenditure when calories are being restricted and by immediately capturing any surplus calories in our diet and locking them down as fat deposits.[23] Screw you, human body! In fact, it appears that exercise alone is often ineffective for weight loss and can sometimes do more harm than good. It stimulates our appetites in proportion to any calories burned.

Why are our bodies so impossible when it comes to weight management? It is because we are built to withstand a life of famine and starvation. For nearly all of our evolutionary history, obesity, heart disease, and weight-related diabetes were essentially nonexistent, so we did not evolve many defenses against them. On the other hand, starvation was a daily threat. Everything about our infrastructure for energy metabolism is reflective of that. In fact, that is exactly why heart disease and obesity are so common now—our metabolisms are poorly adapted for the food climate that we currently live in. The current Western diet is so mismatched to what our bodies were designed to cope with that most of us are "fat, sick, and nearly dead" (the title of a popular documentary on the subject). Depressing, no? Things could be worse. You could be killed by your spouse (see chapter 7).

Just to end things on a slightly more uplifting note, I should remind everyone that healthy weight management is still possible, even with bodies that are seemingly built to thwart us. Fad or crash diets never work over the long term. Hundreds of studies have proven that. Instead, healthy and sustainable eating styles must be adopted permanently. Opt for more fruits and vegetables, fewer desserts and meat. More whole and raw foods, fewer processed and sweetened ones. Cut out the soda and trade out fruit juice for vegetable juice. Combine high-protein and high-fat foods with high-fiber and low-density foods. In other words, you can have a rich main dish, if the portion is small and you have veggies or salad as a side, instead of fries or baked potato. Try to eat more slowly and drink plenty of water as you do. Have whole fruits for dessert instead of cake or ice cream. Treat yourself on occasion, but keep it rare and the portions small. Walk and cycle whenever possible, and take the stairs instead of the elevator every day. Throw in some regular cardiac exercise, and you have a recipe for healthy energy management. It is not always easy, and sacrifices must be made, but most people find that the resulting "high" of feeling healthy and energetic makes the new life pattern easier to maintain. (Or so I hear.)

FURTHER READING

Dawkins, Richard. *The Selfish Gene*. Oxford: Oxford University Press, 2006.

9

AFRAID OF THE DARK

F EAR IS ONE of our most basic emotions. It is the most automatic, instantaneous, and difficult to control. Fear can be traumatic and terrorizing. It can paralyze us or cause us to flee at speeds greater than we would have thought ourselves capable. Animals obviously get scared also. Animals jump when startled and can develop chronic, even irrational fears. With fear and anxiety, our kinship with animals is obvious.

While the simple fear of being startled at jarring and unknown sights or sounds is rather easy to understand, things like phobias and nervous anxiety are the results of more complicated cognitive events in the mammalian brain. By probing the biological roots of our nervous quirks, we highlight our shared ancestry with other animals and shed further light on the evolution of the human mind.

In fact, there is a school of thought in the world of psychology called psychoevolutionary theory that holds that fear is one of a very small number of simple preprogrammed emotions in humans.[1] Along with fear, the other basic emotions are joy, anger, trust, surprise, anticipation, stress, and despair (or sadness). According to this theory, these few so-called primary emotions form a suite of very simple and very ancient emotional programs shared by all or most vertebrates, especially mammals. In humans, more complex emotional states are formed as combinations of the primary emotions, which then interact with memory associations, both conscious and subconscious. The cognitive ability of humans allows further emotional qualia, such as denial or self-concealment, delayed or repressed emotional processing, defensive false memory formation, and on and on. The key point for our discussion here is that underneath the human emotional experience are a few primary emotions, and fear is among them.

These basic emotions are sometimes called primal emotions by biologists and psychologists in order to underscore their savage, untamed nature. Primal emotions represent humans in our most animal-like state. These emotions not only require little or no thinking but also are immune to our attempts to overcome them through reason. Just as you cannot talk yourself out of being hungry, you cannot easily talk yourself out of being afraid.

Charles Darwin himself conducted a self-demonstration of this principle. He went to the reptile house of the London Zoo and pressed his face against the glass, behind which was a feisty snake. He knew full well that the snake would lunge at him, and he also knew that he was perfectly safe. Although he braced himself, determined not to flinch or recoil in fright, he simply could not stop it from happening. In his diary, he wrote of the experience: "My will and reason were powerless against the imagination of a danger which had never been experienced."[2] His conclusion was that neither civilization, education, experience, nor willful effort could easily overcome the fear response built into the animal brain.

Furthermore, the cognitive abilities of humans have augmented and amplified fear in certain situations. This reveals as much about our evolution as it does about the fear programming itself. Read ahead . . . if you dare.

STARTLE FEAR

The simplest kind of fear is startle. This is what happens when an unexpected sound, sight, or feeling is thrust upon us. This fear response is quick and strong. Imagine yourself sitting in a quiet room, reading a really good book—for example, this one—and then suddenly, out of nowhere—bang!

A loud noise interrupts the silence. What will you do? You will startle. You will jump, sit up straight, and either shout, scream, or gasp. You may even drop the book. In addition, your heart rate and blood pressure will skyrocket. Your heart will pump not just faster but also harder, and you will feel it pounding in your chest. Your brain will seem to race, and any sense of relaxation that you were getting from the book will be a distant memory. In short, you will be amped up on adrenaline and experience a classic fight-or-flight response.

Many things frighten us. Imagine walking into your darkened bedroom at night and finding a large cardboard cutout of a vaguely humanoid shape. Your sudden viewing of this unexpected and possibly threatening form will shock you and send you into a fight-or-flight terror response. Sights, sounds, and even touching can induce startle. We all know how fun it is to sneak up on someone from behind and grab his abdomen forcefully from both sides. Wow, will he scream!

Curiously, the startling stimulus does not have to be particularly well disguised as something threatening. For example, you could sneak up on a family member reading quietly and begin to suddenly sing "O Canada" very loudly. There is nothing at all threatening about the song "O Canada" (or about Canada, for that matter), and the person knows very well who you are, and yet, she is briefly terrified. Similarly, the aforementioned cardboard cutout in the darkened room could be a cartoonish figure—something that would register as more funny than scary—and yet, the instantaneous horror still results. Why? If the entire stimulus is easily recognized as familiar and nonthreatening, why are we still so dramatically frightened, even briefly?

The reason we get scared by startling stimuli, even very familiar ones, is because our fear response is much quicker than our recognition response. The "fear program" is tripped when unexpected stimuli enter our brain from our eyes, ears, or skin. Because the program is a short and simple neural circuit, involving few brain regions and few neurons, it is fast.[3] Recognition, on the other hand, requires several rounds of information processing. This involves higher association areas that must carefully compare the sensory input with stored memories until a "match" is found. While this seems to happen very fast on our timescale, it is turtle speed compared with the fight-or-flight response. In the case of startle, you have already jumped ten feet in the air before your brain has put it together that the sudden loud sound was that of your dear husband singing "O Canada."

There are parallels to this. If you have ever accidentally placed your hand on an electric range that you were not aware was hot, you may remember that you actually pulled your hand back before you even felt the pain. This is because the pulling back from a burning stimulus is a prewired automatic reflex. The neural circuit is so short that it involves only a few neurons and is handled by the spinal cord alone; the brain is not even needed. Consciously registering the burning pain, however, involves a longer circuit and takes place in the brain. That takes more time. It is actually pretty amazing when you think about it: you have already reacted before you even feel anything. In fact, part of the startle reaction that you experience when you burn yourself is because of the activation of the reflex arc itself. Some unseen agent has suddenly and forcefully pulled your arm with no warning, like some sort of phantom force.

Both the burning reflex arc and the startle response reveal something profound about the human brain. Some responses are hardwired into the more rudimentary or "primitive" parts of our nervous system. These primitive brain systems are ancient and shared with pretty much all animals with a nervous system, including many invertebrates. They are automatic self-defense mechanisms that operate without conscious control. In the natural world, danger often comes quickly. There is no time to lose. If we were to first take the time to analyze a startling input, recognize what it is, and then evaluate whether or not it actually poses a threat, we might very well get eaten in the meantime. It is the same with the burning stove: the automatic reflex arc saves at least a half second, which makes the difference between a first- and second-degree burn. When it comes to protecting ourselves, it is best to jump first, ask questions later.

This view of startle fear fits very well into the psychoevolutionary theory. Startle is the simple combination of two primary emotions, surprise and fear, and is a very basic and rapid function of our brain. This is the same fast-wired response that allows houseflies to dodge the swatter so swiftly.

It is clumsy, but fast. Even a ringing telephone—hardly a threatening or unfamiliar phenomenon—can give us a fright. The familiarity of the stimulus does not matter because we simply do not have time to recognize it. It is the strength and sudden nature of the stimulus that triggers fright.[4] When we experience a strong, unexpected stimulus, we react defensively and immediately and fire up the sympathetic nervous system (fight-or-flight response); better safe than sorry.

Animals also experience startle. It is a primal reflex to strong, sudden, unexpected stimuli that evolved as a self-defense mechanism against dangerous encroachment. It makes sense that this response is exaggerated in prey animals, like rabbits, and less intense in apex predators.[5] Nevertheless, because startle is so easily recognized as similar, even identical, between humans and animals, we have probably spent enough time on it already.

SUSTAINED FEAR

Complex fears are generated by a completely different kind of neural circuitry than startle fear. Conscious, sustained fear involves higher brain centers and requires substantial processing.

The most common of sustained fears is fear of the unknown. If we all close our eyes and imagine a place that we find particularly scary, we each will probably come up with something different. However, there will be many things in common.

First, we will almost certainly think of some place with which we are not intimately familiar. If you are asked to picture a scary place, you will probably not think of your bedroom. Second, it will almost always be a place that is darkened. Third, what is most scary about the place that we imagine are the things that are not immediately seen or known; the scariness is what might be there, what might happen. The hallmark of sustained fear is that we do not have all the information. Something is hidden; something is lurking.

We feel most comfortable in our own homes because they are familiar. There are no surprises. You can think of many places that are more spacious and more luxurious, but your home is your home partly because it hides no unknown dangers. Similarly, dimly lit places are scary because we cannot see everything clearly. Think about it: a room can seem very scary until you turn on the lights. Then all is well. What changed? Now you can see everything in the room—you have all the information. A darkened room, however, can conceal dangers. The central theme of fear of the unknown is the possibility of hidden danger that could threaten us.

We share fear of the unknown with animals. Anyone with a pet can tell you that. Recently, we took our dog, Bruno, to the apartment of some friends of ours. These friends are no strangers. They have been to our house numerous times, and Bruno absolutely loves them. When they come over, he recognizes them instantly, and his tail wags so hard that his whole back end swings from side to side. These friends also have a dog and have brought her over many times so that Bruno can bond with her. Everyone knows and likes each other. However, when Bruno was first in the new and strange environment of their apartment, he behaved very differently than when the same cast of people and dogs were in his own home. At first, he stayed very close to us. He was on high alert and was constantly sniffing everything in reach. After a while, he slowly ventured out on his own, gingerly tiptoeing into the new territory. He explored the sights, sounds, and smells very cautiously, as if something were lurking there to jump up and bite him.

Fear of the unknown boils down to caution. When we are afraid, we tend to move slowly, keep our eyes open wide, and check all possible sources of danger. Caution is a way of "priming" the body for a flight-or-flight response. By putting our senses, nerves, muscles, and cardiovascular system on high alert, we can react much more quickly to a sudden danger. Similar to the startle response, the caution response has served animals well for hundreds of millions of years because, from time to time, it has saved their lives. If you have ever watched a predator chase its prey on the Discovery Channel, you understand that the difference between life and death can be measured by a hair's breadth. A delay of a few milliseconds can mean that the prey does not get the head start it needs to escape. When there is any possibility of danger, our nervous system goes into "battle mode." Just in case.

It turns out that sustained fear can also be learned and conditioned. In the early days of psychology, the famous Ivan Pavlov showed the effect of conditioning in dogs. Other scientists were eager to test the classical conditioning theory on humans. Unfettered by today's standards of ethics, psychologist John Watson performed fear-conditioning experiments with human children.[6] His hypothesis was that many types of fear are actually learned, not innate. He pointed to families of snake charmers in India in which the children grow up so accustomed to snakes that they never develop any fear of them, while those in most of the West are actually taught to fear snakes as a symbol of danger and even evil.

Watson exposed a particular child, given the pseudonym Albert, to white laboratory rats. Albert showed no fear of them and in fact was delighted by the playful and social rats. (Contrary to common assumption, laboratory rats are gentle and good-natured.) According to Watson, this argued that fear of rats is not innate but learned through their association with the plague, poor hygiene, and vermin infestation. However, Watson went further and trained Albert to fear the rats. Whenever Albert would move to touch one of the white rats, Watson would make terrifyingly loud noises by banging a hammer on a steel rod. Startled, Albert came to associate the fear with the rat, and it did not take long before he was scared of the white rats by themselves.

Given that Albert was not even two years old at this time, his understanding of the world around him was not very sophisticated, and the fear conditioning was not very precise. Albert became frightened of essentially everything white and fluffy. Watson reported that Albert developed a fear of white dogs, white rabbits, and even white coats. At one point, Watson entered the room with a Santa Claus mask complete with a white beard. Albert cried and screamed. Unfortunately, little Albert died at age six, so no long-term follow-up was possible, nor were there the controls and repetitions that are required of modern psychological experimentation. Nevertheless, the point was made that fear can be a learned and conditioned response in humans. Later psychologists demonstrated this with much more rigor.

Animals can also be conditioned to fear. Experiments have shown this in hundreds of different species from snails to monkeys and most commonly rats.[7] Fear conditioning can involve loud noise, electric shock, simulated drowning, or induction of physical pain, and animals can be conditioned to fear virtually any sight, sound, or even smell. Rats have been trained to fear very specific sounds, such as particular songs or even a single musical tone. Pet owners will know well what their animals' fears are. My dog, Bruno, is terrified of just about any new object that I bring home. Tape measures especially freak him out, and he will run away growling any time I take one out. Because we adopted him from the ASPCA, we do not have his full history, but it seems likely that he was harmed or threatened with some handheld object.

Research into conditioned fear in animals has identified the part of the brain where it takes place: the amygdala.[8] This brain region is part of something called the limbic system, which houses many of our primal emotional responses. The limbic system seems to be key for modulating mood, personality, and some of our most basic appetites and behaviors. Some have even called the limbic system the emotion center of our brain.[9] Because the amygdala is at the heart of the limbic system, it seems that fear is at the heart of our emotional lives—at least in terms of anatomy, if not psychology. Not surprisingly, the anatomy of the limbic system, especially the amygdala, is well conserved among all animals. The amygdalae of mice, sheep, chimps, and humans are strikingly similar, when viewed in cross-sections of the brain.

In animals, invasive experiments have confirmed that the amygdala is the "fear center" of the brain. In a blunt example, rats with an injured amygdala will walk right up to cats with no hesitation whatsoever.[10] As the saying goes, the lamb shall lie down with the lion.

Unsurprisingly, PET scans have revealed that the amygdala also mediates fear in humans. While the amygdala and the limbic system are often considered the seat of our primal emotions, there are extensive connections to the "higher" brain areas (e.g., the cerebral cortex). Presumably, this is how fear and other emotions can sometimes be more complex in humans. Our cerebral cortices are dramatically larger than those of other animals, which explains our advanced ability for reasoning, calculation, and other cognitive functions.

For humans, the neural connections between the cerebral cortex and the amygdala allow us to feed advanced reasoning to our brain center. For example, seemingly out of nowhere, we can suddenly have a frightening thought: "I left the stove on!" or "This mole looks like melanoma!" or "A burglar could easily break in through our basement window!" We might even randomly conjure up a thought of someone close to us dying. Leaving aside the question of where these thoughts may come from, once they exist they are automatically fed to the amygdala and we experience the sensation of fear. It does not matter how random or irrational the fear is, at least initially. Fortunately, the connection to our cerebral cortex can also be used to disengage the fear response in the amygdala.[11] To a certain degree, we are capable of self-soothing by thinking through our irrational fears and exposing them for what they are.

This may be an area in which we have a substantial advantage over other animals. Because their cognitive abilities are not as advanced, their ability to reason themselves out of fear is quite limited. As dog trainers and animal handlers of all stripes can tell you, often the only way to alleviate fear in an animal is by generating new associations for the fearful stimulus. The theme is not just confrontation of the fears but also providing new conditioning to supplant the prior conditioning.[12] This is the equivalent of talking oneself out of being afraid except that, in the case of dogs, the calming input does not come from reasoning in the cerebral cortex but through new stimuli that cause a reconditioning of the fear association. (See the discussion on phobias.)

In animals with more sophisticated memories and more advanced cognitive abilities, the emotion of fear can be amplified by our abilities in anticipation of the future. With this great skill (or curse), we can associate past terrors with current situations in more than just a simple stimulus response, like a dog being afraid of a thunderstorm. Anticipation, also called dread, is the fear that something bad is about to happen.[13] Sometimes we get anticipatory anxiety for no reason at all except for past experience.

For example, when things are going well in a relationship or with a project at work, we often think to ourselves that this is usually when things start to go wrong. There is an idiom for this feeling: waiting for the other shoe to drop. As the nineteenth-century French writer Honoré de Balzac famously said, "Our greatest fears lie in anticipation." As proof of this axiom, researchers have recently found that the dreading of future pain is often more unpleasant than the pain itself.[14]

Not all anticipatory fear is pointless dread. Humans have the ability to learn vicariously, which can lead to healthy avoidance fears. We do not have to experience danger firsthand in order to avoid it ourselves. We teach our children about potentially harmful people and things, and we use public service announcements to warn about specific dangerous situations. If we see someone else slip and fall, we will be more careful. If we see someone get shocked by pushing a button, we will not even think of pushing that button ourselves. We are fortunate that we can learn to fear in this way. It has saved countless human lives over the years.

Scientists have actually studied this form of vicarious learning in humans. A group of psychologists from New York University conducted an experiment in which individuals witnessed other individuals being subjected to painful fear conditioning.[15] Human subjects were hit with mildly painful electric shocks following the presentation of an easily recognizable pattern of colored squares on a computer screen. Meanwhile, observers would watch this while having their brain activity imaged. Interestingly, when someone watched someone else being fear conditioned, the brain activity looked strikingly similar to what happened when fear conditioning was experienced firsthand. Even though the observer was in no danger of painful stimulus himself, the amygdala was still activated.

This probably sounds familiar. Earlier, in chapter 3, while discussing empathy and contagious yawning, we learned about mirror neurons and how our brains are capable of taking another's perspective. It seems that the same sort of neural circuitry is at work during the social learning of fear.[16] In fact, it is conceivable that this was the first form of perspective-taking to have evolved, given that fear is much more widespread in the animal kingdom than is empathy. As a means of danger avoidance, "fear contagion" is a very valuable survival tool. By learning from the experience of others, we gain the wisdom without having to endure whatever painful price came with the lesson.

By remembering past dangers, whether first- or secondhand, we learn to avoid them in the future. Similarly, when there is a lack of familiarity, we often fill in the blanks with the possibility of danger. This is why we are cautious in unfamiliar surroundings, shy around new people, and so on. This brings our discussion of sustained fear full circle. The unifying theme of most fear is a fear of the unknown. Caution, anticipatory dread, shyness, and timidity all flow from fear of the unknown, and this fear has served animal species very well. There is usually little cost in being overly cautious and the possibility of great gain by avoiding potentially serious dangers.

PHOBIAS

While healthy fear of real danger is no doubt adaptive, what can be said of phobias? Humans have a somewhat bizarre tendency to harbor persistent irrational fears. Some of these fears are actually somewhat reasonable upon inspection. My former roommate grew up in the desert, where scorpions and spiders are a deadly threat. Accordingly, she has an intense case of arachnophobia. While the intensity of it certainly brings this fear into the realm of the irrational, especially when triggered by harmless daddy longlegs, it developed out of a response to a true and lethal danger. Even when it is extreme, there is little downside in being too scared of spiders.

However, plenty of phobias cannot be understood this way. For example, I have a mild case of emetophobia—the fear of vomiting. As of 2016, it has been thirteen years since I last had the displeasure of vomiting, and since the age of seven or eight, I have only thrown up on two occasions. Despite very rarely throwing up, I frequently obsess about it. With the tiniest rumble in my tummy, I begin to panic. It can interfere with my sleep, and I may refuse to leave the house until the feeling completely subsides. If I am out, I will insist on returning to the comfort of my home because the thought of vomiting in a public place is more than I can tolerate. It can be paralyzing. I have been known to sit in the bathroom for several hours until the nausea passes. One time, I descended into a vomit panic that lasted several days. I did not eat, and I called out sick from work. There was nothing wrong with my stomach—it was nothing more than a phobia-induced anxiety attack.

It is not just my own vomiting; I have a very strong aversion to other people throwing up as well. I know, no one is a big fan of puke. This is more than that. I can obsess over it. If anyone tells me they are not feeling well, I will immediately start assessing the likelihood of that person throwing up, and I will be on high alert. I am quite uncomfortable around sick people, and when someone has had too much to drink, I will do everything I can to keep my distance from them. Just in case. If someone in my subway car looks ill or drunk (which happens a lot in New York City), I will change cars at the next station.

My emetophobia is completely irrational. Throwing up is a natural body process that can be quite protective. By hurling spoiled or infectious material, we help keep ourselves healthy. Plus, the few times that I have thrown up, it was not really that bad. I certainly survived, and I had plenty of warning to get to the bathroom. It is a bit scary, but you feel better afterward. As I constantly ask myself, what is the big deal?

I call this a mild case of emetophobia because it does not actually meet the clinical definition. People with diagnosed emetophobia often avoid children at all costs, eschew social gatherings involving alcohol, and almost never eat at restaurants or other people's homes.[17] They only trust food that they prepare themselves, often with very specific cleaning and cooking rituals, and some will even restrict their diet to a few "trusted" foods. Some emetophobics have developed atypical anorexia in their relentless pursuit to avoid vomiting. None of that applies to me, so I consider myself lucky. I love children, parties, and eating out. However, if a child is sick or someone at the party drinks too much, the anxiety kicks in.

I use emetophobia as an example that resonates with me, but there are thousands of very irrational fears out there. Some phobias, like arachnophobia, acrophobia (fear of heights), claustrophobia, and trypanophobia (fear of needles and other sharps), are easily traced to a very normal fear of a potentially dangerous situation. Psychoanalysts tell us that these kinds of fear usually have their genesis with an incident, often in childhood, of real or perceived danger associated with the object of the phobia. In these cases, a terrifying situation has left a permanent imprint on the amygdala. When that situation is encountered again, the fear program is initiated. As a primal emotion, fear is extremely difficult to control. Overcoming phobias may take years of confrontation, desensitization, reassociation with pleasant experiences, and even psychotherapy.[18]

However, it is more difficult to imagine how a real danger could have caused other phobias. For example, there are people who are afraid of cheese, the color yellow, the number four, beards, telephones, and the sun. There are even people who are afraid of being happy, afraid of someone loving them, and afraid of heaven. These are real, documented phobias that are often debilitating for the sufferers, as you can probably imagine.[19] It is very difficult to see how these particular things place someone in severe danger. However, the usual cause is that some other thing really did threaten or frighten these people, and the fear-conditioned response was misdirected to something loosely associated. For example, maybe a bearded person abused or scared someone, and that fear developed into pogonophobia, a fear of beards in general. Perhaps a terrifying event occurred in a room that was painted yellow. My point here is that these seemingly silly phobias can actually derive from a truly dangerous or terrifying experience. The "fear learning" that took place was appropriate; things just got discombobulated during the programming step.

Do animals get phobias? In a sense, they do. Dogs can develop fears of thunderstorms, mailmen, and household objects that they associate with a past trauma or terror. Loud things like storms and vacuums may have never actually harmed them, but they can still develop a persistent fear of them because of the startle that they have experienced. If the switching on of a loud vacuum or a sudden crack of deafening thunder frightens a young puppy, this will leave an imprint on his amygdala and prime him for fear in the future. It does not matter that he was not actually harmed or in danger. The amygdala processes fear only; it does not consider the follow-up repercussions, nor does it reason its way through the fear and calculate the true danger of the stimulus. If it did, dogs would not be afraid of vacuums and thunderstorms. Human amygdalae are no better. There are humans who are afraid of balloons!

Documented dog phobias include fears of men, children, stairs, fireworks, thunder, noisy appliances, and cars.[20] It is not difficult to see how dogs could develop these fears from something that startled or hurt them in the past. Dogs then get stuck on these phobias because they cannot reason themselves out of them. In fact, each new exposure can actually strengthen the fear, even if nothing happens. This is because the fear response itself is self-reinforcing.

Daisy, the dog I grew up with, was terrified of both thunderstorms and fireworks. At the first sign of either one, she would take off for the basement and stay there the rest of the day. While no harm actually came to her, she was amped up on fear and adrenaline, which feeds the association of the loud sounds with the fear response. For this reason, without intense reassociative therapy, phobias tend to get worse over time.

As alluded to earlier, dogs can overcome phobias with therapy. What we should have done with Daisy was provide new associations for the stimulus of thunderstorms. We should have pulled out her favorite treats and games whenever a thunderstorm started and bombard her with positive stimulus. If necessary, we could have begun in the basement, her safe spot. Although it would have taken consistent repetition to work, this is a proven method of conquering phobias in dogs. It works because it is based on solid biology. By associating storms with treats and games, we could have re-routed that particular stimulus away from the amygdala and toward the pleasure centers of her brain. We should have replaced the prior conditioning with new conditioning.

This is very similar to the way that psychiatrists treat human phobias, supporting the notion that humans and animals process fear in similar ways. In phobia treatment, confrontation is key. The only way to disarm a fear is to face it. Importantly, psychiatrists accompany the confrontation with something pleasurable, comforting, or distracting—the goal being to create new associations in order to dislodge the previous ones. Just like with dogs, it is not one and done. Confrontation therapy takes repeated sessions and serious commitment from the patient, but any psychiatrist will tell you that it can work.[21]

However, humans have an additional curse with some of these phobias. Our advanced cerebral cortex is capable of great imagination. We can create images and ideas in our heads de novo. This is great for our creativity and spatial problem solving, but it is bad if you have a phobia. The mere mention of the feared item can lead to imagery so vivid that the phobic person can experience a full-on panic attack. This is why conquering phobias is more difficult in humans than in dogs.

Interestingly, biologists have discovered that many animal phobias and aversions correspond well with lifestyle and habits.[22] For example, many nocturnal animals are known to exhibit photophobia and actively avoid brightened environments. This behavioral avoidance is based on fear, and the amygdala is involved. This means that these animals are actually afraid of brightness and many also suffer from what zoologists call lunar phobia.[23] I mean that the animals fear the moonlight, not the moon itself. On clear nights when the moon is particularly bright, nocturnal animals are much less active than they normally are and tend to restrict their movements to the moon shadows cast by trees and shrubbery.

Why would nocturnal animals have developed photophobia and lunar phobia? Nocturnality is a strategy employed by prey animals to help them escape predation. By hiding during the day and being active only at night, prey animals make themselves less visible to their predators and thus harder to hunt. Fear of bright moonlight would further that purpose and be a valuable survival tool for nocturnal prey animals. According to this logic, we should not expect to see photophobia among predatory nocturnal animals.

Indeed, that is what we find. Among nocturnal animals, both predators and prey are active at night, but only the prey animals are wired to actually fear the light. In fact, some predator species display what is called lunar philia and are drawn to the moonlight.[24] This is likely an effort to catch glimpses of their prey should they accidentally wander toward the light. (Many other nocturnal predators have evolved keen senses of hearing and even echolocation to track their prey.)

My point here is that animals can and do develop certain aversions to things in their environment that can be harmful to them. The fear that one develops within one's own lifetime is a poor means to avoid dangers because it requires learning; it requires actual exposure to the lethal danger. If fear can be genetically preprogrammed, it will be much more effective. That is indeed what happens. The photophobic and lunar-phobic animals do not learn their fears; they are built in. Long ago, the mutants that were spontaneously averse to light had higher survival rates. Continued predation likely strengthened that genetic fear over time. That is natural selection in action.

Interestingly, photophobia appears in various nocturnal animals and not in their diurnal (active during the daylight hours) cousins. In other words, over the course of evolutionary time, the fear of brightened conditions has likely been turned on and off multiple times. There are several nocturnal prey species among our own order of primates, but they are not all closely related to each other. Most of these species have close relatives that are not nocturnal at all. And yet, among the nighttime prey species, most are photo- and lunar-phobic.[25] Apparently, the "switch" that controls the genetic program for light fearing is easy to turn on and off as a given lineage of animals evolves.

Survival is a constant struggle that periodically forces revolutions in lifestyle. If we were able to see backward through the full family history of just about any mammal—even humans—we would probably observe alternating periods when the ancestors were either nocturnal or diurnal, and maybe even periods when they were metaturnal (being active during both day and night) or crepuscular (being active at dawn and dusk). Based on research into light phobias of extant animals, the fear of light or dark probably played a role in mediating those lifestyle transitions.

FEAR CAN BE GENETICALLY PROGRAMMED

It is widely known that animals can be genetically programmed to fear certain things, such as mice fearing cats and fish fearing birds. Are humans also genetically programmed to have certain fears? Research into some human phobias seems to indicate that we are. I described above how most human phobias derive from a childhood trauma. However, it seems that we are prone to fearing certain things more than others. Have you noticed that fears of spiders and snakes are more common than just about any other fear that you can think of? This may be something that is more than purely learned.

Scientists studying fear conditioning have noticed that it is easier to train humans to fear snakes and spiders than things like friendly dogs and fluffy pillows.[26] This is especially true for children, and indeed, some very young children are afraid of these animals before ever having encountered them or heard about them.[27] There appears to be a preconditioned human tendency to fear snakes and spiders. This would make sense because throughout human evolution, these animals have accounted for countless deaths, and having an inborn avoidance instinct would bring a distinct survival advantage, especially for young children.

Importantly, these studies have revealed that humans have a predisposition toward fear of snakes and spiders rather than a universal and firmly programmed fear of them. Predispositions do not always blossom into reality; they just increase the likelihood thereof. Plenty of people out there actually love snakes and spiders and end up keeping them as pets or studying them as a career. For these weird people, in the absence of a conditioning event, the predisposition toward fearing these animals was never triggered.

If there really is a human predisposition toward fearing snakes and spiders, we might expect to see this in some of our close relatives as well. After all, the lethal danger that these critters pose is in no way restricted to human beings. (Obviously, we would not expect to find it in any animals that have evolved to hunt snakes or spiders since they would have necessarily lost any such aversion.) Studies have shown that virtually all monkey species show a fear of snakes in the wild, while most monkeys in captivity do not.[28] However, this does not address the issue of predisposition. Most humans are not born afraid of snakes, but they are much more likely to become afraid of them than they are to become afraid of most other kinds of animals. The question is, do other primates show a predisposition toward fearing snakes or spiders?

To answer this, Susan Mineka and Michael Cook at Northwestern University conducted a set of clever experiments with rhesus monkeys that revealed that the connection between predisposition and exposure is even more intricate than previously thought. In this experiment, the researchers tried to train naïve monkeys to fear snakes and crocodiles by having them watch videos of other monkeys behaving fearfully toward snakes and crocodiles.[29] It worked! The naïve monkeys actually "caught" the fear of snakes and crocodiles after watching conspecifics being afraid of them. No monkeys were actually harmed, and there was no fear conditioning with loud sounds, shocks, or pain. The observer monkeys simply saw from the sounds and body language of the monkeys in the videos that they were afraid of the snakes and crocodiles. This shows that, in monkeys, the fear of danger can be learned from others, not just through direct experience.

The experiment went further still. The vicarious fear conditioning was not effective when the researchers attempted to train the monkeys to fear flowers. In this setup, the researchers had to use creative splicing and editing to create convincing videos of rhesus monkeys behaving as though they were afraid of some artificial flowers. However, when other monkeys watched this video, they did not acquire any conditioned fear of real or artificial flowers. The fear contagion simply did not work. This fascinating experiment shows that there is a specific predisposition toward fearing snakes and crocodiles in monkeys.

It turns out that most primates fear snakes just as much as most humans do—and for good reason. Throughout the long evolutionary history of primates, snakes have consistently been among their most deadly preda-tors.[30] This means that the human tendency to fear snakes may have been inherited from our primate ancestors.

Anthropologist Lynne Isbell has even made the controversial claim that detection and avoidance of snakes has had a substantial impact on the evolution of primate vision, fear, and intelligence.[31] Her thesis holds that in primates, including humans, one of the main evolutionary forces in the honing of our visual skills was our constant need to spot and identify snakes. Then, we developed the fear and avoidance of those snakes. Finally, natu-ral selection favored those primates that were able to remember where the snakes were, figure out how they hunted, learn to avoid them, and so on. In other words, according to Isbell, one of the strongest forces driving the rapid development of primate intelligence was avoiding and outsmarting snakes.

I might not go that far, but it seems undeniable that fear of snakes is not so much a phobia as an ancient, natural, and justified fear of a lethal pred-ator. I knew it! I agree with Indiana Jones on this one: man, I hate snakes.

ANXIETY

All of this talk about fear naturally leads to a discussion about anxiety. Anxiety is classified as a mood, while fear is an emotion. How are they different? Emotions tend to be very specific and short-lived. They are mea-sured in seconds or minutes and are often reactions to specific stimuli. Moods, on the other hand, tend to be longer lasting and more generalized. While moods are sometimes also traceable to specific stimuli, there are often multiple summative factors at play. Moods typically last hours or days. Fear is when you see a snake in the grass. Anxiety is knowing that you might see snakes on your upcoming camping trip.

Further down that spectrum are personality traits, which are even longer lasting and seemingly impervious to daily stimuli. Anxiety is a mood because it is longer lasting than a momentary emotion but usually not as fixed as a personality trait. Complicating this, someone can be an anxious person as a seemingly permanent part of his personality, meaning that he is often in an anxious mood, appropriately or not. Further still, anxiety can be a disorder of mood in which someone experiences anxiety persistently, inappropriately, and intensely. Although the mood, personality trait, and disorder known as anxiety are three different things, they are united by common brain chemistry.

Anxiety is defined as an unpleasant state of inner turmoil or dread that is focused on an undesirable future event. This dreaded event may or may not be likely to actually happen, but the anxiety comes from the supposition that it could. With this description, anxiety sounds like a mood that must certainly be restricted to humans, no? After all, dread for the future requires the ability to anticipate and contemplate future events. That is a higher-order skill usually reserved for humans.

Dogs have long been known to suffer from separation anxiety. You have probably heard stories of dogs that develop destructive behaviors when left alone. This is attributed to separation anxiety because it is most frequently observed in dogs that have suffered abuse, abandonment, neglect, or other traumatic experiences. As rescued dogs form new attachments and develop trust, the separation anxiety is often eased, if not cured. Our own rescue dog, Bruno, used to chew up his bed when left alone. Although it took time, he finally quit doing it once he began to really trust that we would come back each time that we left. This story is not unique. Most rescue dogs have some separation anxiety as they transition to new homes.

Dogs experience other forms of anxiety as well, such as at the veterinarian's office or when you pull out the dog brush or the hair clippers. My Bruno becomes anxious when we take off his collar or say the word "bath." Dogs display many of the same behavioral hallmarks that humans do when they are experiencing anticipatory dread, such as pacing, whimpering, fixating, restlessness, and nervous tics.

Dog owners, trainers, veterinarians, and behaviorists can easily distinguish anxiety from startle and regular caution. Bruno is not afraid of the hair clippers. He just does not enjoy what we do with them. If we leave them out for a while, as we have, he does not show fear or avoid them.

When he sees them, he does not jump back or stand on edge. Rather, his tail drops, he looks downward, he sulks, and he usually makes a weak attempt to scurry away. Eventually, he will obey our commands to come here, something he definitely does not do when he is truly frightened (as he is by the tape measure, for some odd reason). If what he is experiencing is not dread for the impending haircut, then what is it?

Elephants are gentle giants with a very sophisticated emotional experience, as evidenced by their inclusion in almost every chapter of this book. Elephants have long been thought by their handlers to experience anxiety and dread. As we will see shortly, elephants exhibit the typical suite of anxiety behaviors, including pacing and nervous tics, and they appear quite sensitive emotionally. They take time to trust and then rely heavily on trusted relationships for their psychological well-being. Elephants have long been known to have exceptionally good long-term memory, a fact underscored by their large hippocampus, the part of the brain most associated with memory. It has even been argued that elephants have nightmares following traumatic events.[32]

Elephants are believed to suffer from post-traumatic stress disorder (PTSD), one of the most commonly recognized and well-understood anxiety disorders in humans. Scientists have found that elephants that have witnessed violent deaths of their fellow elephants display adjustment issues and social disturbances for many years following. This has been described several times, but in two recent studies, researchers found lasting emotional trauma in surviving elephants of a herd that had been subjected to a mass killing or "cull." After seeing their herd mates killed, these elephants showed symptoms not unlike those associated with human PTSD: difficulties in coping with change, impaired community attachments and integration into the hierarchy, social withdrawal, and exaggerated or strange responses to startling stimuli.[33]

In addition, traumatized elephants display what are often called stereotyped behaviors. In animals, stereotypy refers to behavioral tics that are rigidly repeated, often for extended lengths of time. Stereotypy is most often noted as a side effect of captivity, and some animals will go so far as to pluck feathers, pull hair, and self-mutilate. Zoologists now recognize these behaviors as telltale signs of anxiety, which is useful for those interested in keeping animals comfortable and properly stimulated while in a captive environment. In elephants, stereotypic behaviors include pacing, rocking or swaying, and swinging of the trunks pointlessly and repeatedly. The fact that wild elephants display stereotypic behaviors following emotional trauma is pretty conclusive evidence that they are suffering from PTSD-related anxiety.

What about our primate cousins? Just a couple of years ago, psychologists and animal behaviorists from George Washington University teamed up to see if they could detect signs of PTSD in chimpanzees using only case reports and field studies that already existed in the scientific literature. The twist? They used only criteria from the fourth edition of the *Diagnostic and Statistical Manual* (DSM-IV). The DSM catalogs psychological disorders and their diagnostic symptoms in humans. Even using this extremely limited approach, the psychologists were able to positively diagnose a number of chimpanzees with PTSD based solely on the symptomology reported in the available literature.[34] PTSD was detected in captive chimps following invasive medical experimentation as well as in wild chimps that had suffered various forms of emotional trauma, such as witnessing the killing of family members.

I find the most convincing evidence of the existence of anxiety in animals is the similarity in the underlying chemistry across species. By now, we have a decent understanding of the hormones and neurotransmitters that mediate human anxiety. As evidence that we have some handle on the chemistry, drugs called anxiolytics have been developed to tweak that biochemistry and are tremendously successful in some cases. Of course, they are not perfectly successful in all cases, as the causes of anxiety are often very complex and the human psyche has a resilient ability to self-injure.

Not surprisingly, the chemistry that we know is responsible for anxiety in humans is also detected in animals and seems to work pretty much the same way. Completing this circle, in laboratory animals, anxiolytics have been shown to relieve symptoms that we commonly associate with anxiety in humans.[35] In fact, new anxiolytic drugs are usually tested in so-called animal models of stress before they are tested in humans. You see, for whatever reason, it is perfectly acceptable for scientists to attribute stress but not anxiety to animals. I find it rather exasperating that many of the same scientists who oppose the idea that animals experience anxiety demand that potential anxiolytics are first shown to be effective in animals before allowing human trials. Exasperating, but not surprising. After all, when it comes to medical research involving animals, there are some uncomfortable consequences to admitting that we are not so different.

SCARY MOVIES AND HAUNTED HOUSES

So far, we have discussed startle, sustained fear, and anxiety in almost strictly negative terms. This is not to say that they are not helpful for the survival of humans and other animals, but rather that the experience of these feelings is generally very negative. Facing one's fear is agonizing, and anxiety is crippling. In short, being frightened is a terrible and stressful feeling that we all hate.

So why are scary movies and haunted houses so popular? What happens when we get scared? Our hearts race and pound harder, our blood pressure rises, our eyes open wide, we gasp for air, and we tense our muscles. We also feel that our brains are on "high alert," and our minds race. This is called the fight-or-flight response and is activated by something called the sympathetic nervous system. (Although the origin of this term is shared with the origin of our concept of sympathy, the sympathetic nervous system has nothing to do with sympathy or empathy. This name is largely a carryover from older uses of the word.) The sympathetic nervous system is wired for fast, whole-body activation, and it saves the lives of animals every day through its rapid and protective actions.

Once tripped by startle, the sympathetic nervous system activates two parallel bundles of neurons, called the chain ganglia, which run along either side of the spinal cord. The chain ganglia then send their nerves throughout the body to places like the eyes, skin, heart, lungs, muscles, and, perhaps most important, the core of the adrenal glands. The adrenal glands sit atop the kidneys (*adrenal* evolved from the Latin description *ad renal*, meaning "near the kidney") and, when activated by the sympathetic nervous system, release hormones called catecholamines. The most powerful catecholamine is adrenaline (usually called epinephrine in the United States), but there are two other major ones, noradrenaline (norepinephrine) and dopamine, as well as several minor ones.

The adrenal gland releases these catecholamines into the bloodstream, where they travel throughout the entire body. Catecholamines then act together with the sympathetic nervous system to activate many tissues, such as skeletal muscle, heart muscle, and nervous tissue. They even temporarily raise your blood sugar and metabolic rate, giving you a burst of energy. Other tissues, including the kidneys, liver, pancreas, and intestines, are actually depressed by catecholamines. The purpose of this is to redirect blood flow and energy resources away from the visceral organs and toward the activated tissues. You can worry about digesting your most recent meal later; for now, you need to address the crisis at hand.

Catecholamines also reach your brain, where they do all kinds of interesting things. They increase alertness, awareness, and excitability, and they even enhance your senses. It is obvious how these enhancements serve us well during an emergency, but they also give us a mild feeling of euphoria. Further, these catecholamines also tickle your brain's pleasure center, making them one of nature's many natural highs. There are receptors for all three major catecholamines (adrenaline, noradrenaline, and dopamine) throughout the brain, and their combined effects are heightened awareness and perception, increased mood and motivation, and general increase in brain activity. Because these are such pleasant things to experience, many people actually seek out experiences that stimulate their sympathetic nervous systems and release catecholamines.

We have actually discussed this earlier, in chapter 1. Part of our natural drive toward exercise, competitive physical sports, and other, nonphysical competitions and games is the activation of the acute stress response. The release of short-term stress hormones, including catecholamines and endorphins, gives us an exciting, amped-up feeling. This not only feels good, but it is also good for us. A burst of healthy stress often has the effect of reducing long-term stress. In chapter 1, I compared this with relieving pressure from a valve. Competitions and games that are nail-bitingly tense or involve physical exertion are fun in the short term and healthy in the long term.

Subjecting ourselves to safe forms of fear is similar. When we watch scary movies or go to haunted houses and deliberately subject ourselves to short-term stress, we unleash a large dose of catecholamines. This actually gives us a euphoric feeling—a high. Of course, there is great variability among persons about how pleasant this form of high is. Some despise this form of stimulation, while others can actually get addicted to it. The same is true of anything addictive. Some try it once and hate it; others are hooked immediately. Furthermore, just as with competitive games and sports, the fear response can push the stress valve and actually provide some relief from accumulated long-term stress. For this reason, subjecting oneself to scary things can be another form of "blowing off steam."

However, there are important differences between frightening things and other forms of thrill-seeking or stressful games. First, the startle response is a pure, strong, and rapid activation of the sympathetic nervous system. That is not the case with competitive stress—at least, not normally. As we discussed, the startle response is so rapid that it is activated and well underway before we have time to recognize or analyze the object of the startle. In fact, many behavioral scientists include fear as a central part of the sympathetic response and use the extended term, the fight-flight-fright response. This means that, during startle, the doses of released hormones are large but short-lived.

Also, the hormones themselves are different, though overlapping, between stressful competition and startle. Startle releases only the classic sympathetic hormones, the catecholamines, while competitive stress also involves serotonin, cannabinoids, and, for physical competitions, endorphins. The point is that startle, exercise, competitive physical sports, and nonphysical competitive games are each variations on a common theme: willful engagement of safe forms of tension gives not only a temporary feeling of euphoria but also a more lasting relief from stress.

There is a whole entertainment genre centered on the theme of being frightened: horror. There are biological benefits to horror movies, television shows, and books, as well as performing stunts and taking other danger-ous risks. On the other hand, people can get addicted to the high that they get from danger. Thrill seekers, stunt artists, and extreme-sport athletes have very high rates of serious injury and accidental death. They know the risks but pursue the adrenaline rush anyway. It is not altogether different from drug addiction. In fact, much of the same brain chemistry is involved.

Furthermore, the thrill we get from safe danger, like scary movies, is different from the thrill we get from real danger. If we are in real danger, sustained fear kicks in, and that is no fun at all. If we are not in real dan-ger, we can simply bask in the catecholamine-induced euphoria, comforted by the fact that it is just a movie (or roller-coaster ride, or book, or haunted house). Stunt artists and thrill seekers have likely become so numb to the catecholamine high that they have to seek increasingly dangerous stimula-tion, not unlike a drug addict needing higher and higher doses. Alcoholics Anonymous, Narcotics Anonymous, and Gamblers Anonymous all exist. Perhaps there should be a Danger Lovers Anonymous?

FURTHER READING

Beck, Aaron T., Gary Emery, and Ruth L. Greenberg. *Anxiety Disorders and Phobias: A Cog-nitive Perspective*. Cambridge, Mass.: Basic Books, 2005.

Öhman, A., and S. Mineka. "Fears, Phobias, and Preparedness: Toward an Evolved Module of Fear and Fear Learning." *Psychological Review* 108, no. 3 (2001): 483.

10

THE RICHNESS OF ANIMAL COMMUNICATION

H UMANS ARE ALWAYS communicating with each other, and our exchanges of information can be incredibly sophisticated. With language, humans have the ability to communicate a breathtaking palette of complexity, nuance, wit, subtlety, and emotion. Considering only the text of a written language, the communication potential of a single sentence is nearly infinite. Add to this inflection, facial expression, body language, emotional and social context, culture and custom, and the possibility of multiple meanings, double entendres, and puns, and our language really is boundless in its potential for communicating ideas between individuals. Such overwhelming complexity seems so far beyond anything that animals are capable of that perhaps comparisons between humans and other animals are just silly.

But then again, maybe not. After all, in other chapters in this book, we have admitted a difference in degree, not of kind. I have not suggested that the pair bonds of swans, wolves, elephants, and humans are all the same. I merely suggest that the pair bonds found in other species bear some of the properties of our own pair bonds. Similarly, I do not claim that cows and cats grieve as deeply and profoundly as humans do, just that they do grieve. The point I am emphasizing throughout is that all or most aspects of the human experience have correlates and precursors—not identical carbon copies—in the animal world.

Regarding communication, the precursors are rather obvious. We all know that animals communicate with one another. However, you may be surprised to know how complex that communication is. The jump from animal modes of communication to human language is not quite as huge and insurmountable as one may think. As we will soon see, many animals use body language, dance, and other nonverbal communication in complex and interesting ways. Other animals communicate with touching, pushing, nudging, stroking, and other forms of physical contact. Some animals have a rich vocabulary of audible words, and some even use particular phrasings of those words. The most advanced animals use all of these modes of communication. The difference between humans and other animals seems huge because of the wealth of communication that is opened up by the last few steps in innovation: grammar, tense, declension, and so on.

I think the ultimate evidence that the language divide between humans and other animals is not that big is found in the extremely short amount of time in which that divide evolved. Oral-auditory communication is widespread throughout birds, mammals, and reptiles, and it is relatively certain that many dinosaurs used it, given the intricate anatomy of their larynxes. Focusing on mammals, nearly all lineages exhibit some kind of audible communication, meaning the ancestor of all mammals almost certainly did also. Over the course of about 225 million years, communication has been growing more complex in many lineages simultaneously, given all of the advantages that come with it. The lineage that would ultimately lead to humans diverged from that of chimpanzees and bonobos six or seven million years ago. In other words, for 97.5 percent of the last 225 million years, our lineage was one and the same with the chimpanzee lineage. Any unique innovations in communication have been developed in a very short amount of time.

In fact, most anthropologists consider the "great leap forward" in human history, when humans began to exhibit culture and form complex societies, to have occurred largely due to the development of complex spoken language. The great leap forward was rather sudden and happened in Africa around fifty thousand years ago. Prior to this, humans were anatomically modern but behaviorally very primitive, probably little different than Homo erectus, Homo habilis, or Neanderthals. Most likely due to the emergence of just a few mutations that made the human cerebral cortex capable of much more advanced cognitive abilities, language evolved very rapidly, and humans became capable of speaking to one another in sophisticated ways.[1] These behaviorally modern humans very quickly spread throughout the globe, replacing their only slightly less intelligent forebears in just a few thousand years. With them came much more complex tools, cave art, rituals, customs, and so on.

Everyone alive today, from Northern Europeans to Australian aborigines to uncontacted hunter-gatherer tribes in the Amazon basin, is a descendent of the first fully modern humans who were able to speak with one another using grammar, diction, and conjugation. These first speakers lived in Africa and migrated out in several waves, overtaking and interbreeding with the more primitive humans that had already colonized the world even as far away as Australia. While we may never know what the first human language sounded like, it is likely most related to the "click languages" of Southern Africa, which contain more linguistic diversity among themselves than is seen between English and Chinese.[2]

The point I am making is that, although our language abilities do indeed seem exponentially more advanced than those of other intelligent animals, it is almost certainly due to only a slight advance in brain function. Furthermore, if premodern humans were like our primate cousins and so many other species of birds and mammals on the planet, they were working with an already very advanced toolkit of communication when the great leap forward occurred. It is a somewhat chilling thought: with the right selective pressures and a little luck, chimpanzees could be just a few thousand years away from learning subjunctive conjugations.

BODY LANGUAGE

Even without the more elaborate things like sign language, gestures, miming, or charades, humans can actually communicate a great deal using only their bodies.[3] Much of this is unintentional, but we can give deliberate signals with our bodies as well. Here are two quick examples of each.

We know when a person is either lost or looking for something just by how they carry themselves. You need not even see his face or hands, yet quickly ascertain what is going on. That is unintentional body language. On the other hand, you can communicate aggression toward someone in various ways, including puffing up your chest and shoulders, lowering your gaze, furrowing your brow, folding your arms, and invading personal space. This is an intentional nonverbal communication given in a language that everyone understands.

Animals also communicate with each other using body language. Other than chemical communication via pheromones and the like, body language was probably the earliest form of communication between animals. In fact, there are many communicative postures in animals that seem to mean more or less the same thing in many different kinds of animals, including humans, pointing to a common linguistic ancestry. Just like the Latin word "duo" morphed into "dos" in Spanish and "deux" in French, forms of animal body language can be inherited and then change and evolve over time. For example, is there any doubt what meaning is intended when your dog licks your face when you are sad? Despite the differences, there is no mistaking the commonality.

First, some examples of animal body language that we have already discussed. Dogs and wolves have a great deal of body language for communicating with each other regarding play, as discussed in chapter 1. When one wolf or dog is about to play-fight with another one, she uses the play signal to indicate the lack of true aggression very clearly: she lowers her front end to the ground like a curtsy, while wagging her tail high in the air. Humans are often confused when we see two dogs wrestling because we hear growls, we see teeth bared, and they look really intense in their grappling. There is no confusion among the dogs, though. Watch carefully, and you will see the frequent display of the play signal, shown most notably by the high wagging tail. Also, the play bites are much weaker than real bites. The jaw muscles of dogs are so strong that they could crush each other's bones if they wanted to. To prevent any serious injuries that could result from a misunderstanding, they have evolved the simple play signal, and they use it often.

We also read in chapter 2 how dogs and wolves use nonverbal communication to apologize for offenses. The submissive/guilty look in dogs, or apology bow in wolves, is unmistakable even to us humans. With my own dog, Bruno, I frequently wrestle and play tug-of-war, and he likes when I hold his toys while he chews on them. In all of these cases, he occasionally nips my hand by accident. In these moments, two things are made clear. First, these accidental bites are many times stronger and more painful than his normal play bites while wrestling. This exemplifies how much dogs weaken their bites when playing. Second, he instantly knows when he has nipped me. He immediately stops playing or chewing and looks at me with a downcast expression. He holds his tail downward, stops panting, and slowly approaches me to lick my nose. It is quite obvious to me that he is letting me know that he did not intend to bite me so hard. It works, too. Who could be mad at him when he does that?

It is interesting that dogs will give friendly licks as part of their apologies and human apologies often involve hugs, handshakes, or kisses. There is even the common expression "kiss and make up." I do not think it is a coincidence that the apology body language employs signals that, in other contexts, indicate attachment and bonding. As you are getting over an argument with a friend or acquaintance, it often ends with a handshake to indicate "no hard feelings." The handshake is a borrowed signal. It is not only for resolving conflicts; it is also used as a friendly greeting between friends or potential friends. By using a friendship signal after an argument, we are attempting to turn a confrontational relationship into a friendly one. Enemies do not share hugs and handshakes; friends do.

It turns out that kissing, licking, and other face-centered touching are pretty universal signs of affection and affiliation.[4] For dogs, cats, and many other small carnivores, licking is common between mates, friends, and children and parents. Horses rub their noses together. Giraffes rub their long necks together. Elephants stroke each other with their trunks and intertwine them. Our close cousins, chimpanzees and bonobos, are known to kiss, lick, stroke, and hug. Bonobos are particularly affectionate and regularly engage in openmouthed kissing during sex. Humans most certainly did not invent the kiss or even the French kiss.

Body language is not only used to indicate affection and apology; it is also used to express more hostile notions like "back off!" When threatened, cats arch their backs, bare their fangs, and flex the erector muscles on their hair shafts, making all their hair puff up. Of course, this simultaneously gives the illusion of the cat being larger than it is, which amplifies the aggressive signal. Dogs show aggression very obviously as well. They adopt a characteristic threat posture, and, just like cats, they snarl, baring their large, sharp teeth. In fact, the display of teeth seems to be a very common aggression signal among carnivores, which makes sense, considering the damage that these animals can do with their sharp teeth and powerful jaws. This signal is even retained in primates. Chimpanzees and gorillas often bare their teeth during aggressive displays.[5]

Interestingly, the bared-teeth signal, when combined with other cues, also indicates happiness in chimpanzees, just like the human smile. To the untrained eye, a chimpanzee smile may look like a snarl, but primatologists can easily distinguish the two, and obviously so can the chimps.[6] This hints at the complexity that is possible when multiple signals are combined, rather like combinations of letters to form different words. To a person who is not literate in English, the words "one" and "eon" look almost the same, even though their actual meanings could not be more different.

Another example of the "back off" signal is the puffing up of the cobra's head when threatened. Sure, she has venom and will use it if she has to, but first, she will try to warn would-be predators that she sees them and is ready for a fight. Many lizards and birds have aggressive display postures that are used less for deterring predators than for intraspecies competitions for mating. Despite popular belief, in most animal species, all-out fighting for access to mates is rather a last resort. Most confrontations are actually settled through nothing more than aggressive displays and posturing. Smaller, slower, or weaker animals almost always back down when they feel outmatched. Natural selection strongly favors this method of competition because actual fighting would lead to avoidable deaths and injuries, even of the eventual winner. Just like with humans, communication is a much better way to resolve disputes than fighting. "Can we just talk this out?"

SIGNALING THEORY

A few prey animals have evolved ways to let their potential predators know that they are dangerous and it might be better to move on and pick another target. The rattlesnake is probably the most striking example of this. Particularly as juveniles, rattlesnakes are the frequent targets of hawks, eagles, crows, raccoons, coyotes, skunks, and even other snakes. With the exception of whip snakes (which have evolved immunity to rattlesnake venom), all of these predators run a very high risk when hunting the very poisonous rattlers. Their only hope of nabbing one without dying in the process is to sneak up on them. When a threat is spotted, rattlers shake their tail and make a very distinct and conspicuous sound. The predators must have quickly learned this signal because they almost always retreat when they hear it.[7]

For some prey species, it would be silly to feign aggression and pretend to be able to fight off a predator. Imagine a mouse trying to convince a cat that he is big and tough and will fight back. However, some prey animals have evolved means to tell their would-be killers, "I see you. I am faster than you. Do not bother trying to chase me because I have a head start and you will just be wasting your energy." Biologists call this signaling theory, and these displays usually involve prey animals engaging in feats of strength to show predators that they are strong, fast, and/or alerted to their presence.[8]

One of the most famous examples of signaling theory is a behavior in gazelles called stotting (also called pronking).[9] The main predator of the gazelle is the cheetah. The cheetah can sprint faster than the gazelle but has less endurance and cannot maintain speed as well while turning. This means that the cheetah must sneak up on the gazelle and make a sudden surprise attack if she is going to catch him. If the gazelle has a head start, the cheetah has no chance. When a gazelle spots a stalking cheetah, he will start jumping very high, straight up in the air. It is a rather remarkable sight. At first blush, this seems kind of stupid. Here is this gazelle being stalked by a cheetah, and when he notices, rather than running away, he makes himself incredibly obvious. However, what happens next demonstrates the purpose of stotting: the cheetah gives up the hunt and walks away. Stotting is the gazelle's way of telling the cheetah that he sees her, he has a head start, and a chase would be futile.

Stotting is often referred to as an "honest" signal because, since the gazelle has to be in good shape to perform it, it is a true display of physical fitness.[10] This is a fascinating example of co-evolution because the gazelle has evolved to perform the signal and the cheetah has evolved the ability to interpret it. Both species benefit from this communication because they have been spared the bother of a fruitless chase. The gazelle is happy to evolve a way to avoid having to outrun the cheetah every time, and the cheetah is happy to evolve a way to reduce her record of unsuccessful hunts. Chases are energy expensive, after all, and they also are very loud and obvious. After a chase, every potential prey animal in the area is suddenly aware of the cheetah. The cheetah gets one shot. If she fails, she may not eat that day.

Of course, the stotting also warns nearby gazelles of a cheetah in the area. That may have been the reason that the behavior first evolved, but we may never know for sure. Further, stotting may be part of the courtship behavior of gazelles. Given its utility in avoiding both predation and in saving energy, stotting seems like as good a display of fitness as any other. Young gazelles are known to playfully stot, which supports the play-as-practice theory described in chapter 1.

The several species of gazelle are not the only animals that stot. Their cloven relatives impalas, antelopes, and wild sheep are all thought to do so as well.[11] Although domestication seems to have diminished stotting in adult sheep, young lambs are prone to periodically engage in bizarre spastic jumping behavior that seems playful and may be the remnant of the stotting instinct. Other forms of pursuit-deterrent signals have been discovered in motmot birds, Eurasian jays, rabbits, mice, curly-tailed lizards, and even guppies.[12]

As I will discuss later, Diana monkeys give certain calls to warn their fellow monkeys of specific kinds of predators. Their vocabulary includes distinct alarm calls for each of their main predators. Interestingly, leopards and eagles, two Diana monkey predators that hunt by surprise attack, are dissuaded from attacking when the monkeys make the calls. Thus, these calls also function as pursuit deterrents for those predators, not unlike stotting in gazelles. However, chimpanzees also hunt Diana monkeys, but they do so through sustained stalking and chase, not surprise attack. Consequently, they are not at all dissuaded by the warning calls.[13] The chimps do not rely on the element of surprise anyway, so they could not care less if the Dianas are aware that they are being hunted. While this Diana monkey alarm system almost certainly evolved as a warning to conspecifics, the predators have "learned" what they mean. For leopards and eagles, the game is up when they have been spotted, and so they give up and move on.

Signaling theory depends on the signals being truthful. What if Diana monkeys made eagle calls randomly, just to protect themselves on the off chance that an eagle was nearby? After a while, the eagles would lose their training and no longer be dissuaded from attacking based on the calls alone. The dishonest Diana monkeys would find themselves to be victims of eagle attacks, possibly even more often than chance alone because the calls might actually attract the eagles. If the cheating behavior were genetic, dishonesty would quickly be bred out of the population, and balance would be restored. Thus, honesty is self-perpetuating in signaling theory.

FACIAL EXPRESSION

Humans also have an impressive ability to communicate things using only our faces. Once again, this can be intentional or unintentional. Our facial expression is almost constantly advertising our current emotional state and can even betray us by revealing feelings that we attempt to hide. We can also add important nuance to verbal communication using our faces. Given the right context, a simple glance at a friend can often communicate a great deal.

Many of the cues and communication that come from our facial expressions are universal among all of the people in our species.[14] This indicates that this form of communication is much more innate than the recent innovation of spoken language. For example, the languages spoken among the hunter-gatherer tribes of New Guinea have absolutely nothing in common with English or any other Indo-European language. (Most of them have nothing in common with each other!) Yet, a smile means exactly the same in the rain forests of New Guinea as it does in the streets of New York City. This speaks to the unconscious, genetic, and inborn nature of facial expressions because if facial expressions were learned, like language, we would expect divergence among the cultures of the world. Instead, we see striking universality. Further proof of this is that people born blind display the same facial expressions as their sighted neighbors. They could not have learned the expressions by seeing them in others. It seems clear that facial expressions and their meanings are almost completely hardwired in the human brain.

Many animal species also communicate with each other using facial expressions. This appears largely limited to mammals for two reasons. First, invertebrates, fish, amphibians, and reptiles, even if they live in communities, show only the simplest hints of interactive social dynamics and cooperation. When it comes to cooperation and communication in the animal world, birds and mammals are the stars of the show. Second, facial expressions require elaborate overlapping musculature in the facial region, which birds do not really have. There are forty-three muscles in the average human face. Many of them are strange, as skeletal muscles go, because they do not connect two bones in order to allow skeletal locomotion. Instead, they are loosely connected to the skull on one end and dermal tissue on the other. In other words, their only purpose is to squish and stretch the skin of the face. This sort of thing does not exist anywhere else in the body. Much of our face musculature exists purely so that we can make facial expressions.

This is why mammals have the monopoly on facial expressions. We are the only animal group that has both the social-cooperative nature to use communication in the first place and the necessary muscles to accomplish that communication using our faces. One might be tempted to think that mammals evolved our elaborate facial muscles for the purpose of communication since that seems like pretty much all we use them for, but it turns out that our face muscles perform an even more basic and essential function: suckling. By studying very old mammal fossils, paleontologists have discovered that a muscle group that was originally located in the throat was co-opted and relocated to the face of early mammals and resulted in much improved suckling from the then-recent mammalian innovation: the breast. With the new facial muscles and the highly social context of nursing, the stage was set for the invention of facial expressions as a means of silent communication by early mammals who must have lived in constant fear of predation by the dominant animals of their day: the dinosaurs.

Even if not the original function of facial muscles, facial expressions almost certainly evolved quickly thereafter, given how universal they are among mammals. Most mammals have just a few basic expressions. However, the diversity and complexity of facial expression really explodes in the primate lineage. None of this diversity, however, would have been possible if earlier mammals had not already invented the concept of facial expression. Just like the first automobiles were not Porsches, the first facial expressions were quite simple in comparison with our impressive modern array. However, keep in mind that, while simple and clunky, the first automobiles were much closer to Porsches than they were to horses.

In 2010, researchers in Montreal discovered that mice display facial expressions when they are in pain.[15] By using mildly painful stimuli, repeating on many mice, and recording their faces with high-speed video cameras, the researchers were able to document and code five facial movements that were almost always elicited when the mice were in pain. They closed their eyes and scrunched the skin around their eyes. Their nose and cheeks bulged outward and backward (toward the eyes), their ears moved backward, and their whiskers twitched. The researchers observed these responses with a variety of painful stimuli, and the responses were proportional to the severity of the pain. None of the pain that was given was in any way centered on their faces, and yet, their faces were reliable indicators of the pain. The researchers were even able to observe that these facial movements were reduced if pain reliever had been administered beforehand.

I am particularly struck by how similar the mouse facial response to pain is to the human one. They grimace! When you factor in the differences in the shape of the mouse face and the human face, the mouse response to pain is almost exactly what we see in humans. When in pain, we squint and scrunch the skin around our eyes; the muscles in our cheeks ball up and move upward; and while we do not have whiskers, we do usually scrunch up our noses. We are separated from mice by about seventy-five million years of evolution, and yet, our pain grimace is remarkably similar. (If you are wondering why these experiments were done, the research group studies pain and is searching for therapies for human sufferers of chronic pain conditions. These scientists are not sadists. Quite the opposite, in fact.)

Since then, the pain grimace has also been discovered in rats, in which it is remarkably similar to that of mice.[16] Pain responses have also been reported in rabbits, and reports are expected soon regarding the pain grimace in lambs, horses, pigs, and rhesus macaques. It seems that many mammals exhibit the well-conserved pain grimace. This grimace is inborn, not learned, and to be sure of that, the original study was done with infant mice. In fact, with the use of 3-D sonograms, human fetuses can be seen grimacing in the womb by twenty-four weeks' gestation. In this case, the fetuses are not in real pain; they are "practicing" various facial expressions as their nervous system takes shape.

The question remains, why is there an involuntary behavioral program wired into the mammalian brain that causes a grimace to be made whenever we are in pain? What purpose does this serve? Does the grimace somehow mitigate the pain, like when we rub our head after bumping it? Does the grimace offer some strategy to avoid the painful stimulus? Or is it just an accidental by-product of neurons that are spontaneously activated by the incoming impulses relaying the pain stimulus? Nope—the sensory perception of pain and the motor activation of facial nerves are housed in totally different parts of the brain.

The only purpose that has been suggested for the pain grimace is that of communication.[17] By grimacing, an animal is requesting help and also giving a signal to fellow conspecifics. *Watch out! This thing hurts!* Assuming that the message is received, this communication has great value for those that can understand the signal because something that is painful is also usually dangerous. In fact, that is how and why pain evolved in the first place: as a means to train animals to avoid things that can cause injury or death. (Pain also immobilizes us after an injury to allow healing.) In nature, pain equals danger, so it stands to reason that social-cooperative animals would evolve a mechanism to communicate pain as a means to signal potential danger. Because we have only just discovered the pain grimace, the communication-of-danger hypothesis has yet to be tested definitively.

However, a different communicative function of the pain grimace has already been proposed and more or less proven. By grimacing, infants communicate their discomfort to their mothers or other caregivers. This is akin to crying but more subtle. In humans, we interpret babies' signs of distress and pain using logic and common sense, but animals have been able to understand and meet the needs of their young long before the evolution of the advanced cognitive abilities that we call common sense. Perhaps the pain grimace plays a role in that. Interestingly, suckling has been shown to relieve pain in infants, even when the pain has nothing to do with hunger per se.[18]

All of this discussion of the communication of an internal sensory state (pain) to others also raises the issue of empathy. It seems very likely that the pain grimace evolved within the context of empathy as the very first means that mammal infants use to communicate with our providers, even before we have developed the motor control to do just about anything else. This possibility has already been raised and demonstrated in humans.[19] The discovery of the pain grimace in many other mammals only emphasizes the likelihood that empathy and emotional communication long predate humanity and were refined over many millions of years of evolution. As we saw in chapter 3, empathy and communication are almost certainly intertwined because empathy *is* communication.

OK, enough about the pain grimace. Primates have seen an explosion in their diversity of facial expressions over the last fifty million years. The facial muscles are mostly conserved across the entire primate phylogeny. The fact that these muscles are only really useful for making faces tells us that facial expressions are likely a universal feature of primates. Indeed, that is what scientists have found. However, the in-context meaning of the individual expressions can vary widely among different primates.

Rhesus macaques have several documented facial expressions.[20] For example, macaques have a fear grimace, which, to us, is indistinguishable from the pain grimace. I expect that the context makes it clear to them. I find it very telling that both pain and fear, which both can indicate potential danger to others, use the same facial expression. This makes great sense because why invent a new signal to communicate the same basic idea?

The macaque also has several different smiles. There is one smile for signaling submission to a dominant conspecific, and there is another for intimidation, which is almost the opposite idea. The difference between the two is based on where the monkey is looking while she gives the grin and whether or not she makes a sound along with it. The position of the tail can also change the meaning of a facial expression, just like with dogs. Lip smacking signals affection and alliance, but it is unclear if this is communicative or merely indicative. Macaques even have an expression for disgust.[21]

Orangutans, which are often solitary and territorial as adults, also use the teeth-baring smile in order to display aggression. Interestingly, eye contact also seems to be a rich form of communications in orangutans.[22] For example, an orangutan who stares at another directly in the eyes is telling him, "Back off if you know what's good for you." On the other hand, affection between allies, mates, or relatives is often indicated by shifty-eyed expressions. Orangutan friends look at each other indirectly, through the corner of their eyes, and quickly look away if their gazes lock. The avoidance of direct eye contact is a sign of trust and friendship. Orangutans have many other facial expressions as well, and never are these shown off more than when juveniles are playing. Despite being mostly solitary as adults, young orangutans are energetic, playful, and highly social. Young orangs will make all sorts of faces at each other and engage in a lot of face mimicry.[23] The meaning of most of these facial expressions has eluded scientists so far.

Gorillas also have many of the primate facial expressions that we have discussed so far, but they have added a few of their own. You can find

YouTube videos of zoo gorillas sticking their tongues out at visitors. Although this may be the result of training or mimicry of humans, the tongue display is indeed part of gorilla body language in the wild. Gorillas will show their tongues when they want to signal that they are done playing or when they simply want to withdraw socially. If one gorilla tries to pull another along in order to play, he might pull back and display his tongue as if to say, "I do not want to play right now." The message usually works, too, as the first gorillas will often then go find some other playmate to solicit. Some researchers have even spotted a young gorilla displaying her tongue after being disciplined by an older gorilla.[24] That seems similar to the equivalent gesture in human children.

Chimpanzees are the animals in which facial expression has been studied in the greatest detail. Chimpanzees have nearly identical facial musculature to humans and consequently are capable of as broad an array of facial expressions. Some of the well-understood facial expressions are the whimper, smile, grimace, hoot, pout, and play face.[25] As noted regarding other animals, the bared-teeth expression can mean several different things, depending on context. Flashing the teeth can signal dominance or threat, but it can also signal submission. It can signal fear or pain, but also indicate happiness, like a smile.

To us, it seems confusing that a single facial expression—bared teeth—can mean so many wildly different things based on context and combination with other simultaneous signals. I doubt it is confusing to the chimps, though. This should not be too hard for us to understand because the exact same thing is true for human smiles. Think about how many different kinds of human smiles there are. There is the calmly pleased smile; there is the elated smile. We smile when we are in love or lust, and we smile when we are being smug or indignant. We often smile when we deliver surprising news or juicy gossip, and we may also flash an angry smile when a rival has screwed us over in some way. There is the smile of mockery and the smile of incredulity. We sometimes smile when frustrated, and we also smile when we hear something fascinating, as if to say, "Huh! Imagine that!"

If you had only a photograph to judge, you may not be able to tell which smile someone was giving. However, in the live-action context, it is almost always obvious, especially if you know the person well. This is why I am confident that chimps have little trouble distinguishing between the different chimp smiles.

The similarity between chimpanzee and human facial expression runs deep. In the 1970s, psychologists developed the Facial Action Coding System (FACS) as a means to apply objective criteria for recording human facial expressions for research purposes.[26] Coding systems like this are crucial for the scientific exploration of phenomena like facial expression because, in order to take careful data that is comparable between different observers and research groups, scientists must find ways to take objective empirical measurements of things that are normally considered subjective. That is what the FACS system does for facial expression. Different scientists working in different parts of the world studying different research subjects can now directly share and compare their data about human facial expressions.

In 2007, however, Dr. Sarah-Jane Vick and colleagues did something quite interesting with the FACS system: they applied it to chimpanzees.[27] What they found was that the chimpanzee facial expressions were so similar to the human ones that the coding system worked wonderfully with only modest adjustments needed based on small differences in the shape of the chimpanzee face. Since that time, scientists have begun using this modified system, called ChimpFACS, to record, code, and communicate with each other about the facial expression of chimpanzees. Armed with a long-awaited method to catalog the expressions, we will have a rich understanding about the facial expressions of chimps. That, in turn, will likely further reveal the emotional features of these, our closest relatives.

Recently, primate researchers in Japan recorded the brain waves of a chimpanzee while showing her pictures of chimpanzees that were either lounging around doing nothing or giving facial expressions or postures that indicate alliance, affection, and affiliation.[28] This was the first electroencephalogram (EEG) of a conscious chimpanzee. The experiment revealed that reactions in the chimpanzee cerebral cortex can be observed immediately (average time: 0.21 seconds) when the subject views pictures of other chimpanzees showing affection, while the "neutral" pictures of chimpanzees giving no clear emotional cues cause little or no higher brain response. The authors of this study humbly noted that the responses that were observed in the EEGs of the chimpanzees are similar to what is seen in the EEG of a human experiencing emotional contagion or empathy.

Putting aside the truly explosive potential discovery of empathy brain waves in chimpanzees using EEG, these results demonstrate that the recognition of facial expressions is hardwired in the chimpanzee brain, just as it is in the human brain. This recognition is lightning fast, automatic, and unconscious. The expressions themselves are strikingly similar between chimps and humans, which is why we are pretty good at understanding each other's faces. The rich diversity of chimpanzee facial expression rivals that of humans, and we probably have not discovered all of them yet.

I want to close this discussion of facial expressions with a bit about our closest companions. Dogs have been evolving alongside us for one hundred thousand years. During that time, their evolution has been subjected to very intense selection due to choices we have made regarding which dogs we allow to breed and which we do not. This is referred to as artificial selection rather than natural selection because it is the result of choices we humans impose on a species instead of the natural destiny that the species would have had without our intervention. Artificial selection is what gave us Chihuahuas, Great Danes, dachshunds, and Old English sheepdogs. It is also what gave us cabbage, broccoli, cauliflower, kale, brussels sprouts, and a few others, all of which came from a wild, cabbage-like plant that humans cultivated and selectively bred during the late Iron Age.

Selective breeding gave us the wide diversity of dog breeds we see today, but long before that, humans used artificial selection to domesticate wolves into companion animals. Wolves that made good companions were favored and allowed to breed. Wolves that did not were killed or released back into the wild. Some of the features that early humans selected were intelligence, obedience, tameness, gentleness, guarding behavior, herding behavior, loyalty, and attachment to humans. By selecting and breeding the "good" wolves and casting aside the "bad" ones, humans were able to domesticate them genetically. We did not have to retrain each generation to be a good companion. We shaped their natural behaviors and instincts into what we wanted them to be. The creation of what we now see in dogs is a marvel of biology and human will.

Whether intentionally or not, this genetic domestication also developed dogs' ability to read, interpret, and respond to the emotions of their human companions. It makes sense. Dogs that would respond to their owners' emotional states would become the "favorites." Imagine a dog that gets happy when his owner is happy, gives comfort when his owner is sad, becomes alert and defensive when his owner is scared or angry, and so on. This dog will be the favorite of his owner, coveted by everyone else in the village. His owner will want to breed him as much as possible for himself and his family and friends. Sympathetic and responsive dogs would be much favored over the aloof mongrels. Over time, this is how dogs became so reactive to human emotion.

We already touched on this, but I raise this issue again here because it turns out that dogs actually have the ability to read human facial expressions. In 2011, researchers trained dogs to discriminate between different photos of their owners: one of the back of his or her head, one of his or her face while smiling, and one of his or her face while giving a blank stare. Most of the dogs were able to quickly learn the task of picking out the picture with the smiling face. The dogs were simply performing a trained task to select the same picture repeatedly. Next, however, the dogs were asked to choose between three new pictures of their owners, again with the owner smiling in one of them. The dogs were able to do it much more often than not. These were different pictures, but the training had taught them to look for the smile, and they were able to do it. They recognized a smile. Finally, these same dogs were asked to find the smile in a group of photos of someone else, a stranger. Once again, they were able to do this much more often than if by chance. Dogs seem to be able to quickly and reliably recognize a human smile, even from a static picture of someone they have never seen before.[29]

That dogs have the innate ability to recognize and pick out a smile from still photos is further evidence of their co-evolution alongside us. However, it also argues that wolves, the ancestors of dogs, already had the neural hardware to recognize facial expressions. It is unlikely that this ability would have emerged in dogs from nothing, as a whole new skill. That is way too much for selective breeding to accomplish in such a short amount of time. Natural and artificial selection work by incremental tweaking, not the invention of brand-new traits. Even the selected evolution of the complex behavioral trait of herding was built on behaviors that the wolves possessed already as pack animals. In the case of smile recognition, dogs rather easily learned to read human faces because they already had the ability to read dog faces.

Interestingly, the ability to read facial expressions works the other way, too. Humans are especially keen at reading the emotions of dogs, possibly more than those of any other animal, including our close relatives whose facial anatomy and expressions are so similar to ours. For example, inexperienced humans often misread the various bared-teeth expressions of chimps and other apes, but we almost never misread the expressions of dogs. A study in 2013 showed that humans could correctly identify the facial expressions of fear, anger, disgust, happiness, sadness, and surprise using only still photos of a dog they had never met.[30] Not every person got every picture correct, but the overall performance was very strong.

Perhaps the most surprising thing revealed by the experiment was that having a great deal of experience with dogs did not help someone identify the facial expressions more accurately. If that had been the case, we could chalk this up to experience and learning. However, the experienced and the inexperienced were equally good at identifying the positive emotions, and people with much experience with dogs actually performed worse in identifying the negative emotions of fear, anger, and disgust. This was probably because their bias toward dogs made them prefer not to think of the dog that way.

What this says to me is that the co-evolution of dogs and humans goes both ways. We selected for certain features in dogs, but they might have influenced our evolution as well. While this may be hard to accept if you picture humanity as we exist now, keep in mind that the domestication of dogs occurred many eons ago, long before civilization and even before the invention of agriculture and the domestication of livestock. In those years, life was a great struggle. Having a trusted companion as a partner in that struggle would have yielded great advantages. Being able to understand that companion would have helped even further.

Is it so far-fetched that the ability to accurately read dog emotions would have provided a survival advantage to early humans? Might it be beneficial to understand what dogs were feeling, for example, when they were alerted to the presence of an intruder or stalking prey? It seems to me that a primitive human who was tuned in to what his dog companion was feeling would have had a distinct advantage over one who was not. This advantage would be further amplified as humans began to cultivate and domesticate other plants and animals. Dogs almost certainly played crucial roles with that process, guarding the stored food, herding the animals, announcing the presence of predators, hunting rats and other pests, acting as a lookout for rival tribes, and so on. Once again, early humans stood to gain a lot by learning to understand their loyal and hardworking companions. It is time to come down off of our evolutionary pedestal and remember that we are not the only organism that has shaped other organisms. We have been shaped as well.

The reason that it was so easy for humans and dogs to co-evolve the ability to understand each other's faces is because the recognition of facial expression was already a feature of both dogs and us. Mammals do not have to be taught how to make or understand facial expressions. We just know. After all, from mice to humans, facial expressions are the quickest, easiest, and most universal way to get one's point across.

VOCAL-AUDITORY COMMUNICATION

While you may not have been shocked to learn that animals engage in various forms of nonverbal communication, you may be surprised to learn that many species actually communicate using words. Webster's Dictionary defines "word" as "a sound or combination of sounds that has a meaning and is spoken or written." By this definition, a large number of species use words. I am not talking about dogs, whose barks can mean all kinds of different things. Barks cannot be considered words because, as far as we can tell, they do not have unambiguous meanings. The same goes for the meow of a cat and the moo of a cow. Those sounds are forms of communication, to be sure, but only derive their meaning when combined with a great deal of other contextual clues. Because they are crude and equivocal, those sounds are not words.

Still, many animals do communicate in a vocal-aural way. While most animal sounds do not quite meet our definition of "words," some clearly do. Everyone knows that birds are prolific singers. All but a few bird species communicate with songs and/or calls. Birdcalls are generally short, discrete chirps or other sounds that are made in response to specific stimuli or needs, while birdsongs are elaborate and impressive tunes that serve their own specific purposes.

Birdsong is a highly gendered behavior in most species. It was once thought that only male birds sang, but female singing has now been discovered in forty different bird species. In fact, in phalaropes, sex roles are exactly the reverse of the common bird paradigm, and the females do all of the singing.[31] Birdsong appears to be used almost exclusively in courtship but can have several different purposes within that context. For example, male song sparrows, one of the most abundant and widespread birds in North America, use songs to attract mates. Because they sing to advertise their maleness and their fitness, instead of by displaying flashy coloring, these birds are a dull brown. Male blackbirds, on the other hand, use songs to establish territory and chase other males away. Seasonally, the blackbird males arrive before the females to establish their respective territories. By the time the females arrive, the singing season is mostly over, and they judge the males by their territory, not their song. In one species of blackbirds, females sing "against" each other as warnings and competition within a harem dominated by a single male.[32]

There are other uses of songs as well. Mountain chickadees sing songs during migration, both during flight and while resting, to help keep the flock together.[33] This is an odd use of songs because most flocks stick together using calls, not songs. Most of us in the northern parts of North America have heard the noisy ruckus of a migrating flock of Canada geese constantly calling to one another, but it turns out that there are a few species that are more melodious about it. Some birds also use songs to scare predators away from a nest or at least let them know that they have been spotted.

The song repertoire of some birds is absolutely astounding.[34] Chickadees and sparrows typically have one species-specific song, and the males compete in their attempt to sing it "better" and thus attract a mate. Ovenbirds have two songs in their "set list" and sing one during the day and the other at night. No one knows why. One species of warbler has ten different songs. He uses some songs for mate attraction and others for territoriality. Scientists have observed more than one hundred different songs from a single mockingbird. Many of those songs are repeated, verbatim, several days apart after singing many other songs. Mockingbirds remember their many songs and will return to them.

The bird with the largest repertoire of songs (that we know of) is the brown thrasher, for which scientists have recorded two thousand different songs from a single bird. However, the thrasher will usually sing each song twice and then move on, never repeating the song again. For that reason, I find the repertoire of the gray catbird to be even more impressive. Although he "only" sings a few hundred songs, he remembers many of them and will return to some of them again and again.

Birds do not always sing their songs alone. Male and female cardinals sing back and forth as part of their courtship. Various kinds of complex duets have been observed in the Hunter's cisticola, an African bird. Sometimes they sing together in perfect unison, sometimes they take turns and attempt to mimic each other, and sometimes they actually harmonize with each other in a coordinated manner.

Although birdcalls are much shorter and less elaborate than birdsongs, it is the calls that are more sophisticated in terms of communicating specific things. For example, for the vast majority of birdsong, the most sophisticated messages that scientists have been able to infer are "please come mate with me" and "stay out—this is my territory." With calls, however, birds seem to be conducting more detailed social communication. Many birds use calls to keep their flock together, as we mentioned with Canada geese. The few bird species that fly at night, rather than resting, are even noisier since they are less able to rely on vision. Birdcalls can also be involved in courtship, of course, as well as territoriality. In other words, calls are used for everything that songs are used for, plus quite a bit more.

Hatchlings and nestlings often make specific calls to indicate hunger, and their parents may make calls to indicate when food is coming. Many bird species can tell their own offspring from others based on their calls. There are even some calls that bird parents use that make their nestlings freeze when a predator is about. It has even been suggested that mallard mothers begin communicating with their offspring before they hatch.[35]

Jays, crows, and other corvids are constantly "cawing" at each other when in social contact for reasons that scientists are still not sure about. Remember that these birds hold funerals for fallen comrades and are quite vocal during the "service" (see chapter 6). During mating season, however, the crows are less friendly and more territorial toward each other and use the caw to chase away competitors. These two kinds of caws sound the same to us, but I doubt they do to the crows. Either context or some auditory inflection surely makes this clear.

Small birds called chaffinches, which are often preyed upon by larger birds such as owls, are known to sound a predator alarm call. First, they use a low-pitched call to let everyone know that a predator is about. Then, once everyone is ready, they switch to a high-pitched call that functions like a battle cry, initiating a coordinated mob attack to chase away the predator.[36]

Interestingly, some animals appear to "eavesdrop" on predator-warning calls made by other species. There is a species of iguana on the island of Madagascar that has evolved very respectable hearing, despite the fact that they do not hunt using sound or communicate among themselves using any auditory communication. They do, however, respond to the predator-warning calls of a species of flycatcher bird that also lives on the island. Both the iguanas and the flycatchers are sometimes preyed upon by large raptors—birds of prey. The iguanas never evolved a warning call for raptors because they did not have to. They could just listen out for the calls made by flycatchers.[37]

Among birds, vocal communication is a nearly universal feature and consists of calls and songs that show a staggering amount of complexity and nuance. Even voiceless birds employ audible forms of communication. Kiwis and some others pound the ground with their feet, storks and albatrosses clap their beaks, woodpeckers drill, and many kinds of birds flap or drum their wings in order to communicate. We are only beginning to understand the complex languages spoken by our feathered friends.

Now, on to mammals. Any discussion of vocal communication among animals would be severely lacking if it did not include mention of whale songs. The songs sung by whales are one of the most fascinating and mysterious forms of vocal communication in the animal world. So haunting and captivating are whale songs that Carl Sagan determined that they be included on the golden record launched into outer space on the *Voyager I* probe. In the summer of 2013, *Voyager I* left the solar system. Thus, whale songs are now in interstellar space (as predicted in *Star Trek IV: The Voyage Home*).

Hundreds of scientists have spent countless hours trying to decipher these enigmatic tunes, and we still only have vague hints as to their function. In fact, for baleen (toothless) whales, scientists are not certain how they make the sounds. In toothed whales (orcas, sperm whales, dolphins, and their relatives), the "songs" are really just high-pitched clicks. Even less is understood about the communication function of these clicks since they are mostly out of our hearing range.

The songs that most people think of when they imagine whale songs are those sung by toothless whales, such as blue whales, gray whales, humpbacks, and right whales. Although they use their larynxes to sing, they do not have vocal cords, so we are not really sure how they do it. The songs are low-pitched and distinctly tonal, and there are a surprising number of parallels between whale song and human music. Like many birdcalls, whale songs seem to operate mostly within what is called equal temperament, and they tend to obey a set key signature.[38]

What are the songs about? Most of the scientific research on whale songs has focused exclusively on their purpose in courtship and mating, but they also help whales find each other, stay together in pods, coordinate for hunting and migration, and so on.[39] This makes perfect sense because life underwater places very different demands on mammals than life on land. Long-range vision is useless, especially in murky or deep waters, and odors diffuse quite slowly in water. Sound, however, travels four times faster in ocean water than in air. So it makes sense that our cetacean cousins moved away from body language and toward vocal communication when they made the transition to aquatic life forty or fifty million years ago.

In humpback whales, perhaps the most prolific singers in the sea, the males do most of the singing while females tend only to make short, scattered sounds that are not organized into recognizable or repeated phrasings. Males sing extensively during mating season. Each song is unique, but a regular pattern of repetition emerges. The notes are organized into phrases, sub-phrases, and songs, which are repeated many times for several hours. Despite concerted efforts to unlock the secrets, scientists continue to debate whether the songs are primarily for attracting females, dissuading competing males, announcing territory, merely staying in contact, or something more complex. Males in proximity with each other will sometimes sing in key with each other and mimic each other's songs, so any attempt to explain the songs as simple female courtship or territorial warnings can be tossed out the window.

Scientists have also detected patterns of regional differences among whale songs of the same species. These "regional dialects" have been discovered in humpbacks and, more recently, blue whales.[40] While the possible function of these differences is still unknown, they have helped researchers in tracking specific populations of whales. By picking out these unique little identifiers, scientists know where the whales are from. It is similar to our ability to spot an Englishman or an Australian by his accent and use of slang.

REFERENTIAL COMMUNICATION

While facial expressions, songs, and calls undoubtedly convey important communications among animals, they do not appear to meet the criterion for being "words" because they do not have clear, unambiguous meanings. In other words, this is not referential communication in which a specific sound or gesture refers to a specific thing or action. Though less concrete and tangible, nonreferential communication is still very important, especially for conveying emotional communication. It also requires sophistication and high-order cognitive skills. However, true referential communication is something else entirely and sets the stage for true language. Therefore, this must only be observed in humans, right?

Hardly. An astonishingly complex catalog of referential communication has been discovered in prairie dogs through thirty-five years of careful observation and experimentation by Professor Constantine Slobodchikoff and his team at Northern Arizona University.[41] Prior to 1980, the "chirps" of prairie dogs were thought to be rudimentary sounds designed to call attention to and also give a general warning about predators. Nondescript predator warning calls are fairly common in both birds and mammals.

Thankfully, he probed further, and this work became a touchstone for a long career in deciphering the meanings of the various chirps. The closer they listened, the more complexity they noticed. Over the next few years, Slobodchikoff and his colleagues discovered that prairie dogs use different chirps for various predators, hawks, bobcats, coyotes, and even humans.[42] Predator warning calls are obviously of great value to prey animals, and having different alarm calls for each predator is particularly useful because the defensive strategy is different for each type of predator. Slobodchikoff noticed that as a prairie dog gave the warning call for a specific species of predator, her fellow dogs took the appropriate evasive action for that predator (discussed later in more detail regarding vervets).

Probing further still, the scientists noticed that the prairie dogs use slightly altered forms of the alarm call for each individual predator animal.[43]

By giving each individual its own "name," the prairie dogs are able to achieve even more specificity and, once a name is given by one dog, the others use that name in the future. Using careful statistical analysis, Slobodchikoff's team discovered that the specific calls that the prairie dogs assign to individual predators are not simply random names: they contain encoded identifying information about the individual. Prairie dogs have descriptive words for the colors, size, and shape of the predators—and even the speed at which they are moving![44]

In addition, prairie dogs have words for things beyond just predators. They make reference to each other and also directional location. Interestingly, their vocabulary is not fixed; they invent new words for novel objects they are faced with, such as cardboard cutouts of various shapes. Slobodchikoff also noticed that the prairie dogs of different regions had recognizably different patterns of chirping, which he called dialects.[45] There is a rough correlation between the complexity of the chirping patterns and the complexity of the local habitat. Further still, regional variance appears proportional to geographic distance, not unlike how human language accents blend, merge, and diverge over geographic distance.

The conclusion of all of this is that prairie dogs have an extensive and adaptable vocabulary for describing certain aspects of their environment to each other. It sounds a lot like language to me.[46]

The work with prairie dogs does not stand alone. Vervets are Old World monkeys residing in the rain forests of Africa. These monkeys also give alarm calls when predators approach. By using recording devices followed by careful observation, scientists have been able to decipher specific calls for certain predators.[47] To signal that an eagle has been spotted, a vervet will make a low-pitched grunt. When a python is spotted, however, a vervet will give a high-pitched, staccato barking sound called a chutter. Finally, when a leopard is spotted, the vervet sings a series of distinct tones.

These are the three main predators of the vervets, so it makes sense that they would call out when they spot one. However, having different calls for each predator is really powerful because each hunts differently, and thus a different escape strategy is necessary for each. If an eagle is stalking, a vervet should seek shade cover or, if in a tree, come down from the top.

If a leopard is hunting, however, it is best to climb up the tree as high as possible. If a snake is close by, better get out of the tree altogether and look for a clearing where the snake can be easily spotted and avoided.

To demonstrate that these vervet calls mean what they thought they meant, researchers performed experiments.[48] They placed loudspeakers in the forest and waited for vervets to approach and then played the various warning calls that they had recorded. Sure enough, playing the eagle call made the vervets look up, playing the leopard call made them climb the nearest tree and head for a thin branch, and playing the python call made them stop and scan the ground around them. The scientists had successfully communicated with the vervets in a way that they naturally understood!

Personally, I think that the most interesting part of this experiment is often left out of the story. It turns out that, when they hear the prerecorded "false alarms," the vervets react somewhat slower and in a more confused manner than when a real alarm call is given. The reason is that, in the real-life scenario, the vervets' first reaction is to look at the vervet that is making the call and ascertain which direction she is facing, which was not possible with the loudspeaker. They do this because vervets always face the threatening animal that they are announcing so that the other vervets can see where the threat is located. If the threat-calling vervet is far off in one direction and is facing away, this indicates that the predator is even further off and that there is no need for immediate panic. The vervets can take their time in seeking safety. However, if the shouting vervet is nearby and looking right in your direction, you had better move your tail quickly because you are in the danger zone!

Although the vocabulary is small, this is a sophisticated form of oral communication that shares many features with our own spoken language.[49] First, the instinct to make these calls is inborn: vervets born in captivity will spontaneously make these various sounds. Second, the proper use and meaning of the calls is learned as young vervets mature. They learn by watching, imitating, and refining, and young vervets that lack adult "teachers" never learn to use the language properly. Accordingly, in the wild, adult vervets do not respond much to the alarm calls of juveniles. This is presumably because they know that the gibberish of young vervets is usually not a cause for concern. Indeed, young vervets make the calls sporadically and inappropriately at first, for example, responding to a leaf falling by giving an alarm call.

As Steven Pinker has powerfully argued, the human tendency to use language is an inborn instinct, not a learned or mimicked behavior, which is then honed through experience and learning as it takes shape.[50] Like human babies, vervets begin to babble and then gradually use the sounds properly by observing adults. The parallel between human language acquisition and vervet alarm calls is striking.

Another interesting thing about vervets is that they can identify their own children by their cries. Animals can tell each other apart by smell and by sight, but only in a few species have scientists been able to observe that individual voices can be discriminated. In the jungle, when an infant vervet cries, his mother looks in the direction of the scream. Meanwhile, all of the other vervets stop and look, but not toward the infant—toward the mother.[51] Not only do vervets know the sounds of their own children, they seem to know which children belong to which mothers.

The three vervet "words" for "eagle," "leopard," and "snake" were discovered by the husband-and-wife team Robert Seyfarth and Dorothy Cheney, who have made primate communication their life's work since the early 1970s. More recently, their study of vervet vocabulary has revealed many more words, including two more predator words, "baboon" and "predator-other." Also, it turns out that vervets have words for more than just predators and dangers. They also have words for social relationships, such as "higher-ranking peer," "lower-ranking peer," and "competing/rival troop of vervets." The differences between these calls are very subtle and will sound like mere grunts to the untrained ear. Only with years of patient listening and analyzing were Seyfarth and Cheney able to decipher this vocabulary and then test their hypotheses using recordings, a testament to the demanding nature of this difficult work.

To date, we have not deciphered any other primate vocabularies as thoroughly as we have the vervets, but this is not because they are anomalous. It is only because no one has yet done with any other species what Seyfarth and Cheney have done with vervets and are now doing with baboons. However, you can be sure that, inspired by this success, primatologists are now doing exactly that with many other species. Soon we will have a more complete vocabulary for many more of our primate cousins.

In marmosets, for example, scientists have observed that small groups of monkeys will often gather and apparently converse. They make a host of seemingly random sounds but with clear repetition of some "words." The most fascinating part of this is that the individual monkeys take turns and politely wait while others are speaking.[52] The dominance hierarchy sometimes rears its head, but everyone usually gets a chance to say something if they want to. The parallel to human conversational etiquette is striking.

Oddly, some scientists do not seem to be willing to characterize these "chat sessions" as conversations because they are operating under the assumption that the vocalizations are largely meaningless. However, I would bet one hundred dollars that if Seyfarth and Cheney had chosen, nearly four decades ago, to work on marmoset communication instead of vervet communication, we would have discovered some of the lexical meaning of the so-called random vocalizations. Further, some scientists are currently arguing that turn-taking in these chatty marmosets may be evidence that human language could have evolved directly from vocalizations instead of from gesturing, as is widely believed. That may be, but to my mind, that fact underscores even more the point that marmosets would have no reason to take turns politely if they were talking gibberish.

Moving on to the great apes, scientists have identified at least seventeen and as many as twenty-five distinct vocal calls used by gorillas. The precise meanings are still being debated because context seems to be crucial for the translation, but precise diction has been documented around the topics of food, fights, dominance, playing, and sex.[53] Fascinatingly, scientists have found that individual troops often develop some novel words or, more often, novel variations of an existing word. This is the gorilla equivalent of regional slang, which, in human language, represents the beginnings of language divergence. I find it hilariously telling that the most well-documented example of this "regional slang" is a variation of the "sex request" among a group of western lowland gorillas. Is that always the case? Sexual slang develops and changes swiftly in human populations, so why not it in gorillas as well?

Although chimpanzees are our closest relatives and possibly the smartest nonhuman animals on Earth, scientists have not been able to document as extensive a vocal vocabulary in chimps as in vervets, baboons, and gorillas. This is because chimps tend to use gestures, touching, displays, and other forms of body language in their communication. However, there is at least one clear and distinct call in chimpanzees that is unambiguous, and that is the pant-hoot, which is a display signal and a call for attention, first documented by Jane Goodall. Interestingly, the chimps have personal variations on the pant-hoot so that other chimpanzees know who is calling. As you might guess, parents respond to their children very quickly, and response rates drop among more distant relationships.

Although scientists have not observed great lexical depth in chimpanzee calls, this is not to say that they do not communicate vocally. Quite the opposite: chimpanzees are incredibly vocal. It is just that they do not use a great many different kinds of sounds, and the ones that they use seem to be rather unrefined, which is to say that they can mean any number of things depending on context. Chimpanzee calls are mostly about getting each other's attention or expressing emotional states. Chimpanzee calls are like interjections: *Wow! Hey! Stop! Over here! Darn!* Things like that. Once a chimp has her friend's attention, she uses nonverbal communication to get whatever is bothering her off her chest.

SYMBOLIC GESTURES IN PRIMATES

Although we have already discussed some forms of body language, gestures are quite a special phenomenon because, by definition, they cross a clear line into something that we have unequivocally observed in only a few animals: intentional representational communication. With most animal body language discussed previously, it is difficult or impossible to know if the actions are the result of an intentional effort at communication or merely reflective of the internal state of the animal. The latter does not mean that they are not communicative. Clearly, the meaning of certain body movements is apparent, and evolution has sprung up around these movements to refine them and strengthen their comprehension. However, it is impossible to know if the signals are given in an intentional effort to say something.

Gestures, on the other hand, are defined as intentional acts of communication performed by the body. They are highly representational. To the extent possible, gestures may mimic or pantomime the idea being expressed. Gestures, like calls, are highly sophisticated forms of communication because they involve abstraction: the connection of two otherwise unrelated concepts. For example, vervet monkeys connect a specific call sound to the nearby presence of snakes. Similarly, gestures will connect a specific body movement to some other concept. This involves very complicated neural circuitry that integrates vision, association, memory, and understanding. As such, intentional gestures have traditionally been thought only to exist in the great apes, the family of primates that includes humans, both chimpanzee species, gorillas, and orangutans.

Beginning with gorillas, we find one of the most famous animal gestures of all time: chest beating. What does it mean? Silverback gorillas beat their chests when challenged. Remember that gorillas live in harems dominated by one—and only one—silverback male. Any time that a silverback feels that his dominance is being threatened, he beats his chest to say "back off." The most exaggerated chest pounding occurs when a silverback spots a rival silverback. This is an attempt to show strength and dominance in the hopes that fighting can be avoided. Gorillas also beat their chests when they feel the need to assert their dominance in other circumstances.

Moving to the chimpanzee genus, zookeepers at the Leipzig Zoo in Germany recently discovered that the bonobos there communicate the concept of "no" by shaking their heads from side to side exactly as humans do.[54] It is possible, perhaps even likely, that the bonobos picked this up from their human caretakers. However, the fascinating part, no matter the genesis, is that the Leipzig bonobos now use this gesture to communicate with each other. This is most often seen when the bonobo mothers will not allow their children do something or want them to stop doing something. It seems bonobo children have the same experience we humans do while growing up—a whole lot of "no."

This is not the only time that seemingly human gestures have been observed in our close chimpanzee relatives. A 2007 study extensively documented the gestural communication of two troops of common chimpanzees and two troops of bonobos.[55] This study documented thirty-one distinct body gestures and eighteen facial-vocal expressions. Once again, the similarity to human gestures was striking. An outstretched hand with palm facing upward forms a "beg" gesture in both chimp species. A "silent pout face" was also discovered.

One thing about chimp gestures is that they are almost always combinatorial, as with facial expressions. Most gestures are combined with posture, facial expression, or even sounds, and the timing and social context is important, too. In other words, it is complicated. I suspect that this is the reason that we are only now starting to decipher ape gestures and other animals' body language. Until now, we could not figure out any simple meanings after watching them for a while, so we just jumped to the conclusion that the animals were too simple for gestures. It turns out that exactly the opposite is true: animal communication is complex, and we were being too simple in our attempts to understand it.

Mark Laidre has proposed that one community of mandrills, the huge African monkeys with brightly colored faces, has developed a gesture for "leave me alone."[56] The gesture is covering the eyes with one hand and holding for an extended period. By watching closely, Laidre observed that the mandrills made the gesture when off by themselves and when they did so, their peers tended to approach and touch them much less often (compared to when they were in the same posture but without the face covered). This gesture may work mechanistically through the prevention of eye contact, which is often how social contact initiates among primates, including humans. This is the first discovery of an intentional gesture in a non-ape species in the wild. In addition, because the gesture is unique to one specific community and stable over time, it bears the mark of cultural transmission. The mandrills learn this gesture and what it means.

It is clear that apes, and possibly monkeys, use gestures to communicate their requests, emotions, and commands. This is not so different from humans, as we all know. In addition, the similarities in many of those gestures are striking as well. Some examples are making the "ask-beg" gesture, sticking out the tongue, shaking the head no, and clapping the hands to indicate elation and excitement. The facial expressions are hauntingly similar as well, and the tone and strength of various screams and calls usually increase in ways that humans would not find surprising. Chimpanzees, gorillas, and other apes even laugh when tickled.[57]

One reason that language has commonly been thought to exist only in humans is our tendency to equate language with spoken language. The oral and pharyngeal anatomy of apes (other than humans) makes speaking, as we know it, impossible.[58] The throat is too shallow and the larynx too high. They simply cannot make many of the sounds that we can. However, there is no such anatomical impediment to the use of gestures in other apes. As we should know from the various forms of sign language in use around the world, language need not be spoken.

The definitive proof that apes are physically and cognitively capable of communicating with gestures came when scientists began to teach sign language to apes. We all know the story of Koko, the gorilla who uses sign language. Koko understands two thousand English words and communicates back with more than one thousand signs. While her signs contain no grammar—a key feature of human language—she has formed novel signs by combining others. For example, she invented a sign for "ring" by combining the sign for "finger" and "bracelet."[59]

Koko's use of signs often drifts into seemingly incomprehensible randomness and repetition, and some skeptics have pointed to this as evidence that her use of signs represents little more than an impressive memory and operant conditioning. I am not convinced. After all, the gestures she is using did not emerge through a natural process of biological evolution over the course of many thousands of years. All of these signs were taught to her de novo, and I think it is unreasonable to expect her to be perfect at using them. Koko expresses herself spontaneously and enjoys "chat sessions," in which she will just sit and converse for a while. There are many videos of Koko on YouTube, including one of her tickling and being tickled by the late Robin Williams. Go watch some of them, and ask yourself if she appears to be repeating meaningless motions in order to get a reward.

A bonobo named Kanzi has mastered even more English words than Koko. He uses a lexigram board and appears to grasp the connections between words and their organization into coherent phrases.[60] For example, researchers asked Kanzi to "make the snake bite the dog" (referring to his stuffed animals). This is a concept that Kanzi had never heard or communicated before, yet sure enough, he retrieved the stuffed snake and pretended that it was biting the stuffed dog. That is a huge conceptual leap that many thought no animals were capable of. Kanzi does not just memorize words; he draws novel conceptual connections between them. Even if we are never able to convincingly document this happening in the wild, Kanzi proves that bonobos have the cerebral hardwiring for this ability.

The most fascinating part of the Kanzi story is that he was not the original "student" of the attempted instruction with the lexigram board; his adoptive mother, Matata, was, but he often accompanied her to her training sessions. To the researchers' great surprise, one time while Matata was away, Kanzi picked up the board and began using it correctly and coherently.[61] It remains to be seen if this was because he is especially gifted or if his being younger made the difference, reminiscent of how deftly human children pick up second languages.

Anecdotally, Kanzi may have also provided evidence that bonobos use some form of vocal language among themselves. One time, he was in a room, visually separated from his sister, Panbanisha, but she could hear him. He was given some yogurt, and he began to make some vocalizations as he ate it. Panbanisha then pointed to the lexigram symbol for "yogurt."[62] Because yogurt is obviously not a bonobo food in the wild, they obviously do not have any inborn native concept of it. In light of that, how did Kanzi communicate this concept to his sister? Had they developed a "word" for it previously, since it is treat they enjoy? Or was this just a crazy coincidence?

It is important to keep in mind that both Kanzi and Koko learned their respective lexica of words through the concerted and determined effort of their human handlers. The concepts were exhaustively repeated in a highly artificial laboratory environment. Many have worried that such experiments in sign language may say more about ape memory, and possibly intelligence, than they do about language abilities. This is because they do not mimic how any species, including humans, learn language. This is a fair point.

Meet Washoe. Washoe was a chimpanzee that was raised, from age two to five, in a house with a human family and treated as any other child would have been.[63] The parents, Allen and Beatrix Gardner, shunned vocalizations around Washoe and used only American Sign Language (ASL) with her. Washoe eventually learned around three hundred fifty words of ASL. Initially, the Gardners had to use concerted efforts to teach Washoe her first signs, but nothing like the laboratory efforts behind Koko's and Kanzi's learning. Rather, their efforts were not at all dissimilar to human parents teaching their human children their first words. The learning of signs eventually became automatic. Washoe would first recognize a sign and react appropriately and then begin to use the sign herself.

Like Koko and Kanzi, Washoe could combine signs in novel ways and express her own independent ideas. Washoe might also have been self-aware. When handed a mirror, she held it for a while, made faces, and explored it curiously. When asked what she was looking at, she responded, "Me." Her awareness also extended to the emotional state of others. One of her favorite human caretakers, Kat, had a miscarriage and missed work for several weeks. When Kat returned, Washoe pouted and expressed her anger or disgust at Kat for abandoning her. This is what happened next: Kat made her apologies to Washoe, then decided to tell her the truth, signing "My baby died." Washoe stared at her, then looked down. She finally peered into Kat's eyes again and carefully signed "Cry," touching her cheek and drawing her finger down the path a tear would make on a human (chimpanzees do not shed tears).[64]

Later in Washoe's life, she actually taught some ASL signs to other chimpanzees, particularly a young male named Loulis that grew up with

Washoe as a role model or adopted mother figure.[65] Loulis used his first sign after just eight days with Washoe and eventually learned dozens more. To ensure that he was not learning the signs from the human handlers, they would use only a few basic signs around him. He learned the rest from Washoe in what can accurately be called the beginning of a self-perpetuating language.

Washoe's environment was much more "natural" than Koko's or Kanzi's—but only in human terms. While it may replicate how modern human children begin to use language, it does not come close to mimicking chimpanzee life. However, that misses the point. The experiment was not designed to ask if chimpanzees learn sign language naturally in the wild. The point was to learn if chimpanzees have the necessary brainpower for the learning of representational language. It seems rather obvious that they do. This means that the innate ability to learn representational language and basic phrase structure long predated the spark that pushed us toward human language a mere fifty thousand years ago.

Most of the work on gestures in primates has taken place in laboratory or zoo environments with subjects that were held captive for all or most of their lives. It is fair to say that this limits how much we can generalize the work to the natural behaviors of the species in question. However, that is not really relevant because the work was designed to see what these species are capable of, not necessarily what they really do in the wild. However, plenty of work on gestural communication among primates in the wild has been done as well. For one, Mark Laidre's study of the "leave me alone" gesture in mandrills was conducted by observing wild mandrills.

In 2011, Catherine Hobaiter and Richard Byrne reported their exhaustive effort to catalog the repertoire of intentional gestures of chimpanzees in the wild.[66] After observing and analyzing well over four thousand instances of gesturing over the course of nearly nine months, they were able to clearly identify sixty-six distinct gestures with apparently consistent in-context meanings. Nearly all of the gestures had previously been reported among some primate, either formally or informally, and every single gesture was documented in more than one chimp. No chimp displayed her own unique gestures.

Fascinatingly, Hobaiter and Byrne noted slight but significant differences in the personal repertoire of gestures based on age (but not gender). This parallels human speech. My grandmother and I spoke mutually intelligible English, but we did not always use the same words. I call it a "couch"; she called it a "davenport." I called it "homework"; she called it "lessons." I called it "hanging out"; she called it "visiting." Americans can read and understand the Declaration of Independence, but we would not use the same words if we wrote the document today. There are generational differences in word usage that accumulate over time in human populations. Perhaps it is the same for chimpanzees. Further, it could be that, as younger chimpanzees mature, their gestural repertoires evolve into the patterns of their elders. That, too, would be interesting, but for different reasons.

Perhaps the most surprising result of this study, besides the large size of the natural gestural repertoire of chimps (sixty-six), was how universal the gestures were. As I said before, many of the gestures the researchers cataloged had been previously seen in some other primate species. In fact, only a few truly novel gestures were reported. The rest had been observed in chimpanzees held in laboratories or zoos. Interestingly, of the sixty-six recorded gestures, most of them had also been reported in one of the other ape species, such as gorillas, orangutans, or bonobos. More than one-third of the gestures (twenty-four) had been reported in all of these ape species, which is a striking amount of conservation.

Because the study did not directly address this issue, the authors stopped short of saying whether they thought the gestures were inborn or learned. The surprising degree of universality argues that there is at least some genetic influence in the gestures. How else would these gestures remain so strictly conserved among species separated by millions of years of evolution? Think of it this way: while some human languages share cognate words because they are recently related, others have almost nothing in common. The English and Aztec languages have essentially zero cognates despite the fact that the speakers of those languages are only divided by a mere thirty or forty thousand years of separate ancestry. The ancestors of orangutans and chimpanzees diverged from one another at least ten million years ago, and yet they retain a third of their "words" in common. Genetics!

Although not explicitly mentioned in the study, I would bet that many of the gestures that are shared among all the great apes are also found in humans. Gestures are an interesting feature of human communication because they, too, are incredibly universal around the world despite the fact that they are not explicitly taught to us as children. We just kind of pick them up.

You might argue that we learn these common gestures simply by watching our elders, and of course that happens also. However, something deeper is probably at work. First, if human gestures are simply learned from others, like spoken language, why are so many of them pan-cultural, while spoken languages are not? Second, if gestures are learned through mimicry, why do people who are born completely blind use gestures? Studies have shown that blind children begin to use gestures even before they learn to speak, just like sighted children.[67]

Blind children and adults also use a great deal of gestures when they speak, and the gestures are pretty much the same as those that sighted people use. Similarly, sighted people use almost as many gestures when speaking to a room full of blind people as they do when speaking to other sighted people. I think these results point to a deep root for our tendency to make gestures. Gesturing is not learned; it is inborn. Experience and culture merely refine our gestures. We "learn" to make gestures the same way that our heart "learns" to pump blood: from our genes.

From the song of a sparrow to the neigh of a horse, from a vervet calling to a woodpecker drilling, from a gorilla beating his chest to a dog wagging her tail, animals have evolved ways to communicate with each other. Through the eons, communication has grown increasingly more sophisticated. Through sight, sound, and touch, animals have learned to reveal their emotions, intentions, and frustrations. Further, referential communication is being discovered in a growing list of mammals. In the journey from mere pheromones to the sonnets of Shakespeare, most of the road was well behind us before we began shooting the breeze in the savannahs of Africa.

FURTHER READING

Halloran, Andrew R. *The Song of the Ape: Understanding the Languages of Chimpanzees.* New York: Macmillan, 2012.

Slobodchikoff, Constantine. *Chasing Doctor Dolittle: Learning the Language of Animals.* New York: Macmillan, 2012.

EPILOGUE

Metacognition, Self-awareness, and the Mind

G IVEN THE DEPTH of animal emotion and the complexity of their communication, it is fitting to close this book with some speculation regarding animal consciousness. Do animals think? Are they self-aware? Do they recognize their own mental states? Do they have minds? These questions are incredibly difficult to answer but critically important for informing our policies toward them.

Just because animals are not able to report their subjective experience to us does not mean that they do not have one. In fact, just because a fellow human can describe his perceptions, they are not any less subjective. We have no way of knowing if other humans experience pain or joy the same way we do, even though we may use the same words.[1] As children, we learned from others about how to describe our mental experiences, so of course we put common labels on them. That does not mean that the inner experience is identical. We do not even know if the perception of the color orange is the same for each of us.

As with their emotional states, the only way we can attempt to get at the issue of animal minds is by observing their behavior. Because this final section is more speculative than other sections, I have chosen only a few experiments that are most interesting to me, but there is a rich body of research in these areas and whole books dedicated to these difficult but important questions.

Western scrub jays, the species of corvids discussed in chapter 6 as holding "funerals" for dead comrades, are known to be highly intelligent and capable of making and executing elaborate plans, including hiding excess food for a later meal. In 2014, Arii Watanabe and Nicola Clayton of the University of Cambridge published the results of an experiment that demonstrates how advanced the cognitive abilities of scrub jays are.[2]

In the experiment, the jays were allowed to look on as researchers hid a piece of food in one of four possible cups. However, sometimes three of the four cups were closed with a large cover, making it obvious which cup was open and thus contained the food (closed cups were never opened and could thus never contain the food). If all the cups were open, however, and the jay did not see, or forgot, which cup hid the food, they would have to check each one until they found it. The researchers first trained the jays a few times by having them observe the two different kinds of food hiding and then letting them retrieve the food.

Then, the researchers performed a simple test. The jays were placed behind a temporary barrier with two peepholes. Behind one of them, a researcher hid food in one of four open cups. Behind the other, another researcher hid the food in one open cup with three closed ones nearby. Because the two researchers were hiding the food simultaneously, the jays had to choose which peephole to look through and thus which researcher to watch.

The researchers found that all five birds that were tested "passed" the test, meaning that they spent their time watching the researcher with the four open cups, rather than the one with three closed ones and only one open. Although this may seem simple, this experiment shows that the jays were able to use experience to make a decision in the present about an event in the future. By observing the cups, they knew that in one set, it would be obvious where the food would be placed—they did not need to actually watch. In the other set, however, there was value in watching the researcher as she placed the food because it would save time later.

The researchers went so far as to claim this was proof of metacognition in corvids because the birds, in making a deliberate choice about what to observe, recognize that they do not have certain information that they would like to have. Awareness of one's mental states, in this case knowledge or lack of knowledge, is indeed the key feature of metacognition. The jays appear to "know that they will know" where the food is hidden in the scenario where the food could only be in one cup. They also "know that they will not know" where the food will be placed when all four cups are open, and so they choose to watch.

Rats also appear to be aware of what they do and do not know. Researchers at Indiana State University trained rats to learn a variety of tests involving successful discrimination of short versus long durations of time (less or more than four seconds).[3] Successful completion of the tests earned the rats six pellets of food, while unsuccessful completion earned them nothing. However, the rats were not always forced to take the tests. They could also decline to take a test and earn three pellets.

When the researchers trained rats to discriminate whistles that lasted two seconds from those that lasted eight, most of the rats would choose to take the test every time. This is because the rats performed well on these low-difficulty tests. However, when other rats were trained to discriminate between 3.6 and 4.4 seconds, a test they often failed, they declined an average of half of the time. In other words, when the rats knew that the task was more difficult and that they had a good chance of not performing well, they elected for the easy reward of three pellets. But when the task was easy, they went for the big payoff.

In addition, the researchers also compared the success rates of rats that voluntarily chose to take the tests with the rats that were forced. The rats that chose to take the test performed substantially better, on average, than those that were not given the option to decline. This means that not only did the rats have some sense of how difficult a test was and their likelihood of performing well, but also that those perceptions were accurate. Again, this argues for metacognitive abilities in the rat, an awareness of one's mental contents.

Moving on to primates, Robert Hampton, then at the National Institutes of Health, reported that rhesus macaques are also capable of "knowing that they know" something.[4] In this experimental setup, the macaques were allowed to choose or decline a memory test based on an image that they had been shown some time before. When forced to take the test, the animals forgot the image about half of the time. However, if given the choice to decline the test for a smaller reward, those that chose to take the test anyway completed it successfully almost every time. This argues that the macaques were aware of whether or not they had forgotten the image. If they knew that they had forgotten, they declined the test. If they knew that they remembered, they took the test and were successful. This phenomenon is sometimes called uncertainty monitoring and has been discovered in all of the apes, some monkeys, and dolphins, as well as possibly elephants.

Hampton's research group, now at Emory University, has also discovered that rhesus macaques have the ability to recall, not just recognize, shapes.[5] When they are shown partial versions of shapes that they have been trained to recognize and asked to complete them, they are able to do so fairly accurately. This is not a trivial distinction because recognition and recall are quite different neurologically. Recognition simply involves comparison of a current stimulus with a stored perception, establishing a "match." Recall, however, involves the recreation of an image in our mind's eye, like when you close your eyes and imagine a house, car, or person that you know.

That monkeys have the ability to conjure images, not just recognize them, implies that the perceptions of their inner experience are more detailed than we previously thought. In addition, it argues that this form of bottom-up visual processing (mental conjuring of images) is quite old in the primate lineage. While this is not metacognition, per se, it is an important feature of the mind and thoughts as we currently understand them. This is particularly important when considering animal minds because, in humans, the mind is intricately linked with language. Because it is debatable whether or not animals truly have language, the other aspects of the mind and thoughts, such as memories, imagery, and metacognition, are crucial.

The fact that nonhuman animals do not use language (as we know it) to communicate with each other has long been the linchpin for the belief that animals do not think and thus do not have true consciousness.[6] However, this line of reasoning has been severely, if not mortally, weakened. First of all, even scientists who agree that thoughts are intimately linked with language do not claim that thought is impossible without language or that thinking is strictly constrained by language.[7] We have all had the experience of not being able to identify just the right word to express something we think or feel, settling for the best approximate. Clearly, our inner experience is richer than our language can convey. If thinking is more than language, it is at least possible that language-less animals might engage in it.

In addition, as demonstrated in chapter 10, animals certainly do have some of the most important aspects of language. For example, many animal species clearly use referential communication and convey novel ideas. Koko, Kanzi, and Washoe have all proven able to express their observations, desires, confusion, fears, and regrets. Their communication is hardly the stuff of Shakespeare, but, well, neither is mine.

It is true that animal communication seems to consist only of nouns and simple verbs without tense, mood, or conjugation, and this is why it falls far short of what we consider language. However, even though human communication usually has grammar and is expressed as complete sentences, human thoughts do not and are not. We formulate ideas and words in our heads in a fairly unstructured way, probably not that different from the communication we observe in other primates and in human children.

A final aspect of the issue of animal minds, and one that many consider to be the most important, is the notion of self-awareness. Do animals have a sense of themselves as distinct from other individuals and within the context of their environment and circumstances? One of the simplest tests of self-awareness is the mirror test, which probes whether animals can perceive a reflection of themselves as such.[8] The test often involves the placement of a mark or item on the animal's body in a place that is hard for her to see without the mirror and then observing if the animal understands that the mark is on her own body and whether she explores or tries to remove it. Using this test, it is has been shown that dolphins, orcas, magpies, elephants, orangutans, bonobos, and common chimpanzees do recognize reflections of themselves (though some of these studies have produced conflicting results).

Some scientists believe that the mirror test underreports the number of animal species that are self-aware because it relies solely on visual cues.[9]

Many animal species, such as rats and dogs, use olfactory sensation to discriminate individuals much more than they use vision. Other species may use aural features. When these animals fail the mirror test, it may not necessarily reveal anything about whether or not they are self-aware. There may be other limitations to the mirror test as well. With something as complex as self-awareness, can we really expect that a single experimental technique can possibly capture it?

For the question, "Do animals have minds?" we have a considerable amount of evidence that the answer is yes. Think about what happens in our minds. We formulate words; we conjure images of things, places, and people; we experience our emotions; and we contemplate our mental contents. We have evidence that animals use words, conjure images, experience emotions, and have some awareness of their mental contents. It is increasingly difficult to defend the position that animals are mindless when they appear to have all or most of the features that comprise the mind.

Fittingly, I shall give the last word on this topic, and of this book, to Charles Darwin himself: "Nevertheless, the difference in mind between man and the higher animals, great as it is, certainly is one of degree and not of kind."

FURTHER READING

Bekoff, Marc. *Minding Animals: Awareness, Emotions, and Heart.* New York: Oxford University Press, 2002.

Griffin, Donald R. *Animal Minds: Beyond Cognition to Consciousness.* Chicago: University of Chicago Press, 2013.

Safina, Carl. *Beyond Words: What Animals Think and Feel.* New York: Henry Holt and Company, 2015.

Smith, Julie A., and Robert W. Mitchell. *Experiencing Animal Minds: An Anthology of Animal-Human Encounters.* New York: Columbia University Press, 2012.

NOTES

INTRODUCTION

1. Steven Pinker and Ray Jackendoff, "The Faculty of Language: What's Special about It?," *Cognition* 95, no. 2 (2005): 201–236.
2. Julie Ann Smith and Robert W. Mitchell, *Experiencing Animal Minds: An Anthology of Animal-Human Encounters* (New York: Columbia University Press, 2012).
3. Matt Ridley, *The Red Queen: Sex and the Evolution of Human Nature* (London: Penguin, 1994).
4. Robert Axelrod and William Donald Hamilton, "The Evolution of Cooperation," *Science* 211, no. 4489 (1981): 1390–1396.
5. Martin A. Nowak, "Five Rules for the Evolution of Cooperation," *Science* 314, no. 5805 (2006): 1560–1563.
6. Joseph Henrich, Jean Ensminger, Richard McElreath, Abigail Barr, Clark Barrett, Alexander Bolyanatz, Juan Camilo Cardenas, Michael Gurven, Edwins Gwako, and Natalie Henrich, "Markets, Religion, Community Size, and the Evolution of Fairness and Punishment," *Science* 327, no. 5972 (2010): 1480–1484.
7. Lynn Margulis and Dorion Sagan, *Microcosmos: Four Billion Years of Evolution from Our Microbial Ancestors* (Berkeley: University of California Press, 1986).
8. Pinker and Jackendoff, "The Faculty of Language."
9. Y. E. Stuart , T. S. Campbell, P. A. Hohenlohe, R. G. Reynolds, L. J. Revell, and J. B. Losos, "Rapid Evolution of a Native Species Following Invasion by a Congener," *Science* 346, no. 6208 (2014): 463–466.
10. Ibid.
11. Chris Stringer, "Evolution: What Makes a Modern Human," *Nature* 485, no. 7396 (2012): 33–35.

1. WHY DO WE PLAY?

1. M. B. Willis, "Genetic Aspects of Dog Behaviour with Particular Reference to Working Ability," in *The Domestic Dog: Its Evolution, Behaviour, and Interactions with People*, ed. James Serpell (Cambridge: Cambridge University Press, 1995), 51–64.

2. Dorit Karla Haubenhofer and Sylvia Kirchengast, "Physiological Arousal for Companion Dogs Working with Their Owners in Animal-assisted Activities and Animal-assisted Therapy," *Journal of Applied Animal Welfare Science* 9, no. 2 (2006): 165–172.

3. Gordon M. Burghardt, *The Genesis of Animal Play: Testing the Limits* (Cambridge, Mass.: The MIT Press, 2005).

4 Ibid., 139–153.

5. Satoshi Ikemoto and Jaak Panksepp, "The Role of Nucleus Accumbens Dopamine in Motivated Behavior: A Unifying Interpretation with Special Reference to Reward-seeking," *Brain Research Reviews* 31, no. 1 (1999): 6–41.

6. Siri Leknes and Irene Tracey, "A Common Neurobiology for Pain and Pleasure," *Nature Reviews Neuroscience* 9, no. 4 (2008): 314–320.

7. Marc Bekoff, "Social Play in Coyotes, Wolves, and Dogs," *BioScience* no. 4 (1974): 225–230.

8. Marc Bekoff, "Play Signals as Punctuation: The Structure of Social Play in Canids," *Behaviour* 132, no. 5 (1995): 419–429.

9. John W. S. Bradshaw and Helen M. R. Nott, "Social and Communication Behaviour of Companion Dogs," in *The Domestic Dog: Its Evolution, Behaviour and Interactions with People*, ed. James Serpell (Cambridge: Cambridge University Press, 1995), 115–130.

10. Marc Bekoff and Jessica Pierce. *Wild Justice: The Moral Lives of Animals*. Chicago: University of Chicago Press, 2009.

11. Bekoff, "Social Play in Coyotes, Wolves, and Dogs," 225–230.

12. K. Groos, *The Play of Animals* (New York: D. Appleton, 1898).

13. Frank E. Poirier and Euclid O. Smith, "Socializing Functions of Primate Play," *American Zoologist* 14, no. 1 (1974): 275–287.

14. Jane B. Lancaster, "Play-mothering: The Relations between Juvenile Females and Young Infants among Free-ranging Vervet Monkeys *Cevcopithecus aethiops*," *Folia Primatologica* 15, no. 3–4 (1971): 161–182.

15. Johanna H. Meijer and Yuri Robbers, "Wheel Running in the Wild," *Proceedings of the Royal Society B: Biological Sciences* 281, no. 1786 (July 7, 2014).

16. Marc Bekoff, "Social Play Behaviour: Cooperation, Fairness, Trust, and the Evolution of Morality," *Journal of Consciousness Studies* 8, no. 2 (2001): 81–90.

17. Anne Pusey and Marisa Wolf, "Inbreeding Avoidance in Animals," *Trends in Ecology & Evolution* 11, no. 5 (1996): 201–206.

18. Anne E. Pusey, "Inbreeding Avoidance in Chimpanzees," *Animal Behaviour* 28, no. 2 (1980): 543–552.

19. Michael Potegal and Dorothy Einon, "Aggressive Behaviors in Adult Rats Deprived of Playfighting Experience as Juveniles," *Developmental Psychobiology* 22, no. 2 (1989): 159–172.

20. Elisabetta Palagi, "Play at Work: Revisiting Data Focusing on Chimpanzees (Pan trog-lodytes)," *Journal of Anthropological Science* 85 (2007): 63–81.

21. J. I. Webster Marketon and R. Glaser, "Stress Hormones and Immune Function," *Cellular Immunology* 252, no. 1–2 (2008): 16–26.

22. William R. Lovallo, *Stress and Health: Biological and Psychological Interactions* (New York: Sage, 2005).

23. Nikolaas Tinbergen, "Ethology and Stress Diseases," *Science* 185, no. 4145 (1974): 20–27.

24. Marek Spinka, Ruth C. Newberry, and Marc Bekoff, "Mammalian Play: Training for the Unexpected," *Quarterly Review of Biology* 76, no. 2 (2001): 141–168.

25. Michael J. Meaney, John B. Mitchell, David H Aitken, Seema Bhatnagar, Shari R Bodnoff, Linda J Iny, and Alain Sarrieau, "The Effects of Neonatal Handling on the Development of the Adrenocortical Response to Stress: Implications for Neuropathology and Cognitive Deficits in Later Life," *Psychoneuroendocrinology* 16, no. 1 (1991): 85–103.

26. Potegal and Einon, "Aggressive Behaviors in Adult Rats Deprived of Playfighting Experience as Juveniles," 159–172.

27. Miriam Schneider and Michael Koch, "Deficient Social and Play Behavior in Juvenile and Adult Rats after Neonatal Cortical Lesion: Effects of Chronic Pubertal Cannabinoid Treatment," *Neuropsychopharmacology* 30, no. 5 (2004): 944–957; Sergio M. Pellis, Vivien C. Pellis, and Ian Q. Whishaw, "The Role of the Cortex in Play Fighting by Rats: Developmental and Evolutionary Implications," *Brain, Behavior and Evolution* 39, no. 5 (1992): 270–284.

28. Pellis, Pellis, and Whishaw, "The Role of the Cortex in Play Fighting by Rats," 270–284.

29. John A. Byers, *Animal Play: Evolutionary, Comparative and Ecological Perspectives* (Cambridge: Cambridge University Press, 1998).

30. Joe L. Frost, Sue Clark Wortham, and Robert Stuart Reifel, *Play and Child Development* (Upper Saddle River, N.J.: Pearson/Merrill Prentice Hall, 2008).

31. Dorothy G. Singer, *The House of Make-Believe: Children's Play and the Developing Imagination* (Cambridge, Mass.: Harvard University Press, 1992).

32. James Paul Gee, *What Video Games Have to Teach Us About Learning and Literacy* (New York: St. Martin's Press, 2014); Kaveri Subrahmanyam and Patricia M Greenfield, "Effect of Video Game Practice on Spatial Skills in Girls and Boys," *Journal of Applied Developmental Psychology* 15, no. 1 (1994): 13–32.

33. Lloyd P. Rieber, "Seriously Considering Play: Designing Interactive Learning Environments Based on the Blending of Microworlds, Simulations, and Games," *Educational Technology Research and Development* 44, no. 2 (1996): 43–58.

34. Ibid.

35. Scott Nunes, Eva-Maria Muecke, Lesley T. Lancaster, Nathan A. Miller, Marie A. Mueller, Jennifer Muelhaus, and Lina Castro, "Functions and Consequences of Play Behaviour in Juvenile Belding's Ground Squirrels," *Animal Behaviour* 68, no. 1 (2004): 27–37.

36. Robert Fagen and Johanna Fagen, "Play Behaviour and Multi-year Juvenile Survival in Free-ranging Brown Bears, Ursus arctos," *Evolutionary Ecology Research* 11 (2009): 1–15.

37. Elissa Z. Cameron, Wayne L. Linklater, Kevin J. Stafford, and Edward O. Minot, "Maternal Investment Results in Better Foal COndition through Increased Play Behaviour in Horses," *Animal Behaviour* 76, no. 5 (2008): 1511–1518.

38. Stuart L. Brown, *Play: How It Shapes the Brain, Opens the Imagination, and Invigorates the Soul* (New York: Avery, 2009), 126.

39. Ibid., 73.

2. ANIMAL SYSTEMS OF JUSTICE

1. S. F. Brosnan, C. Talbot, M. Ahlgren, S. P. Lambeth, and S. J. Schapiro, "Mechanisms Underlying Responses to Inequitable Outcomes in Chimpanzees, Pan troglodytes," *Animal Behaviour* 79, no. 6: 1229–1237.

2. S. F. Brosnan and F. B. M. de Waal, "Monkeys Reject Unequal Pay," *Nature* 425, no. 6955 (2003): 297–299.

3. Darby Proctor, Rebecca A. Williamson, Frans B. M. de Waal, and Sarah F. Brosnan, "Chimpanzees Play the Ultimatum Game," *Proceedings of the National Academy of Sciences* 110, no. 6 (February 5, 2013): 2070–2075.

4. Friederike Range, Lisa Horn, Zsófia Viranyi, and Ludwig Huber, "The Absence of Reward Induces Inequity Aversion in Dogs," *Proceedings of the National Academy of Sciences* 106, no. 1 (2009): 340–345.

5. Sarah F. Brosnan and Frans B. M. de Waal, "Evolution of Responses to (un) Fairness," *Science* 346, no. 6207 (2014): 314.

6. Robert Boyd and Peter J. Richerson, "Punishment Allows the Evolution of Cooperation (or Anything Else) in Sizable Groups," *Ethology and Sociobiology* 13, no. 3 (1992): 171–195.

7. Victoria J. Horner, Devyn Carter, Malini Suchak, and Frans B. M. de Waal, "Spontaneous Prosocial Choice by Chimpanzees," *Proceedings of the National Academy of Sciences* 108, no. 33 (August 16, 2011): 13847–13851.

8. K. V. Thompson, "Self Assessment in Juvenile Play," in *Animal Play: Evolutionary, Comparative, and Ecological Perspectives*, ed. John A. Byers (New York: Cambridge University Press, 1998), 183–204.

9. Ibid.

10. Marc Bekoff, "Social Play and Play-soliciting by Infant Canids," *American Zoologist* 14, no. 1 (1974): 323.

11. Marc Bekoff, "Play Signals as Punctuation: The Structure of Social Play in Canids," *Behaviour* 132, no. 5–6 (1995): 5–6.

12. Tim H. Clutton-Brock and Geoffrey A. Parker, "Punishment in Animal Societies," *Nature* 373, no. 6511 (1995): 209–216.

13. Marc Bekoff and J. Pierce, "The Ethical Dog," *Scientific American* (February 2010), http://www.scientificamerican.com/article/the-ethical-dog/.

14. Milanda Petrů, Marek Spinka, Veronika Charvátová, and Stanislav Lhota, "Revisiting Play Elements and Self-handicapping in Play: A Comparative Ethogram of Five Old World Monkey Species," *Journal of Comparative Psychology* 123, no. 3 (2009): 250.

15. Jessica C. Flack, Lisa A. Jeannotte, and Frans de Waal, "Play Signaling and the Perception of Social Rules by Juvenile Chimpanzees (Pan troglodytes)," *Journal of Comparative Psychology* 118, no. 2 (2004): 149.

16. Gerald S. Wilkinson, "Reciprocal Food Sharing in the Vampire Bat," *Nature* 308, no. 5955 (1984): 181–184.

17. Meredith P. Crawford, *The Cooperative Solving of Problems by Young Chimpanzees* (Baltimore: Johns Hopkins Press, 1937).

18. Ibid.

19. Maynard J. Smith, "The Evolution of Alarm Calls," *American Naturalist* 99, no. 904 (1965): 59–63.

20. Robert M. Seyfarth and Dorothy L. Cheney, "Grooming, Alliances and Reciprocal Altruism in Vervet Monkeys," *Nature* 308, no. 5959 (1984): 541–543.

21. C. J. O. Harrison, "Allopreening as Agonistic Behaviour," *Behaviour* 24, no. 3 (1965): 161–208.

22. Geoffrey W. Potts, "The Ethology of Crenilabrus Melanocercus, with Notes on Cleaning Symbiosis," *Journal of the Marine Biological Association of the United Kingdom* 48, no. 2 (1968): 279–293.

3. MORAL ANIMALS

1. R. M. Church, "Emotional Reactions of Rats to the Pain of Others," *Journal of Comparative and Physiological Psychology* 52, no. 2 (1959): 132.

2. Jules H. Masserman, Stanley Wechkin, and William Terris, "'Altruistic' Behavior in Rhesus Monkeys," *The American Journal of Psychiatry* 121, no. 6 (1964): 584–585.

3. Inbal Ben-Ami Bartal, Jean Decety, and Peggy Mason, "Empathy and Pro-social Behavior in Rats," *Science* 334, no. 6061 (2011): 1427–1430.

4. Nobuya Sato, Ling Tan, Kazushi Tate, and Maya Okada, "Rats Demonstrate Helping Behavior toward a Soaked Conspecific," *Animal Cognition* (May 2015): 1–9.

5. Carolyn Zahn-Waxler, Barbara Hollenbeck, and Marian Radke-Yarrow. "The Origins of Empathy and Altruism," in *Advances in Animal Welfare Science 1984*, ed. M. W. Fox and Linda Mickley (Boston: Martinus Nijhoff Publishers, 1985), 21–41.

6. Marc Bekoff, "Are You Feeling What I'm Feeling?," *New Scientist* 194, no. 2605 (2007): 42–47.

7. H. Markowitz and V. J. Stevens, *Behavior of Captive Wild Animals* (Chicago: Nelson-Hall, 1978).

8. Elizabeth V. Lonsdorf, "Chimpanzee Mind, Behavior, and Conservation," in *The Mind of the Chimpanzee: Ecological and Experimental Perspectives*, ed. Elizabeth Lonsdorf, Stephen R Ross, and Tetsurō Matsuzawa (Chicago: University of Chicago Press, 2010), 361.

9. Lawrence Anthony and Graham Spence, *The Elephant Whisperer: My Life with the Herd in the African Wild* (New York: Macmillan, 2009).

10. S. C. Minta, K. A. Minta, and D. F. Lott, "Hunting Associations between Badgers (Taxidea taxus) and Coyotes (Canis latrans)," *Journal of Mammalogy* 73 no. 4 (1992): 814–820.

11. I. Karpulus, R. Szlep, and M. Tsurnamal, "Associative Behavior of the Fish Cryptocentrus cryptocentrus (Gobiidae) and the Pistol Shrimp Alpheus djiboutensis (Alpheidae) in Artificial Burrows," *Marine Biology* 15, no. 2 (1972): 95–104.

12. "Beluga Whale 'Saves' Diver," *The Telegraph*, July 29, 2009.

13. Douglas Allchin, "The Evolution of Morality," *Evolution: Education and Outreach* 2, no. 4 (2009): 590–601.

14. Jeheskel Shoshani, *Elephants: Majestic Creatures of the Wild* (Emmaus, Penn.: Rodale Press, 1992).

15. Robert R. Provine, "Yawning: The Yawn Is Primal, Unstoppable and Contagious, Revealing the Evolutionary and Neural Basis of Empathy and Unconscious Behavior," *American Scientist* 93 no. 6 (2005): 532–539.

16. Steven M. Platek, Samuel R. Critton, Thomas E. Myers, and Gordon G. Gallup, "Contagious Yawning: The Role of Self-awareness and Mental State Attribution," *Cognitive Brain Research* 17, no. 2 (2003): 223–227.

17. Atsushi Senju, Makiko Maeda, Yukiko Kikuchi, Toshikazu Hasegawa, Yoshikuni Tojo, and Hiroo Osanai, "Absence of Contagious Yawning in Children with Autism Spectrum Disorder," *Biology letters* 3, no. 6 (2007): 706–708; Helene Haker and Wulf Rössler, "Empathy in Schizophrenia: Impaired Resonance," *European Archives of Psychiatry and Clinical Neuroscience* 259, no. 6 (2009): 352–361.

18. Steven M. Platek, Feroze B. Mohamed, and Gordon G. Gallup, "Contagious Yawning and the Brain," *Cognitive Brain Research* 23, no. 2 (2005): 448–452.

19. Ramiro M. Joly-Mascheroni, Atsushi Senju, and Alex J. Shepherd, "Dogs Catch Human Yawns," *Biology Letters* 4, no. 5 (2008): 446–448.

20. E. Palagi, A. Leone, G. Mancini, and P. F. Ferrari, "Contagious Yawning in Gelada Baboons as a Possible Expression of Empathy," *Proceedings of the National Academy of Sciences* 106, no. 46 (2009): 19262–19267.

21. Giacomo Rizzolatti, Luciano Fadiga, Vittorio Gallese, and Leonardo Fogassi, "Premotor Cortex and the Recognition of Motor Actions," *Cognitive Brain Research* 3, no. 2 (1996): 131–141.

22. Evelyne Kohler, Christian Keysers, M. Alessandra Umilta, Leonardo Fogassi, Vittorio Gallese, and Giacomo Rizzolatti, "Hearing Sounds, Understanding Actions: Action Representation in Mirror Neurons," *Science* 297, no. 5582 (2002): 846–848.

23. Evelyne Gallese, "The Shared Manifold Hypothesis: From Mirror Neurons to Empathy," *Journal of Consciousness Studies* 8, no. 5–7 (2001): 5–7.

24. Lindsay M. Oberman, Edward M. Hubbard, Joseph P. McCleery, Eric L. Altschuler, Vilayanur S. Ramachandran, and Jaime A. Pineda, "EEG Evidence for Mirror Neuron Dysfunction in Autism Spectrum Disorders," *Cognitive Brain Research* 24, no. 2 (2005): 190–198.

25. Steven M. Platek, Feroze B. Mohamed, and Gordon G. Gallup Jr., "Contagious Yawning and the Brain," *Cognitive Brain Research* 23, no. 2 (2005): 448–452.

26. Marco Iacoboni, "Imitation, Empathy, and Mirror Neurons," *Annual Review of Psychology* 60 (2009): 653–670.

27. Oberman et al., "EEG Evidence for Mirror Neuron Dysfunction in Autism Spectrum Disorders," 190–198.

28. Shirley Fecteau, Alvaro Pascual-Leone, and Hugo Théoret, "Psychopathy and the Mirror Neuron System: Preliminary Findings from a Non-psychiatric Sample," *Psychiatry Research* 160, no. 2 (2008): 137–144.

29. Scott O. Lilienfeld, Jonathan Gershon, Marshall Duke, Lori Marino, and Frans de Waal, "A Preliminary Investigation of the Construct of Psychopathic Personality (Psychopathy) in Chimpanzees (Pan troglodytes)," *Journal of Comparative Psychology* 113, no. 4 (1999): 365.

30. Frans de Waal, *Primates and Philosophers: How Morality Evolved* (Princeton, N.J.: Princeton University Press, 2009).

31. Charles Darwin, *The Descent of Man: And Selection in Relation to Sex*, 2nd ed. (New York: D. Appleton, 1882).

32. Marc Bekoff, *Minding Animals: Awareness, Emotions, and Heart* (New York: Oxford University Press, 2002).

4. SEXUAL POLITICS

1. Hans Kruuk, *The Spotted Hyena: A Study of Predation and Social Behavior* (Chicago: University of Chicago Press, 1972).

2. Gaël Trionnaire, Jim Hardie, Stéphanie Jaubert-Possamai, Jean-Christophe Simon, and Denis Tagu, "Shifting from Clonal to Sexual Reproduction in Aphids: Physiological and Developmental Aspects," *Biology of the Cell* 100, no. 8 (2008): 441–451.

3. Scott Pitnick, "Investment in Testes and the Cost of Making Long Sperm in Drosophila," *American Naturalist* 148, no. 1 (1996): 57–80.

4. J. Roughgarden, *Evolution's Rainbow: Diversity, Gender, and Sexuality in Nature and People* (Berkeley: University of California Press, 2009).

5. Ibid.

6. Frans B. M. De Waal, "Bonobo Sex and Society," *Scientific American* 272, no. 3 (1995): 82–88.

7. Amy Randall Parish, "Sex and Food Control in the 'Uncommon Chimpanzee': How Bonobo Females Overcome a Phylogenetic Legacy of Male Dominance," *Ethology and Sociobiology* 15, no. 3 (1994): 157–179.

8. Craig B. Stanford, "The Social Behavior of Chimpanzees and Bonobos: Empirical Evidence and Shifting Assumptions 1," *Current Anthropology* 39, no. 4 (1998): 399–420.

9. Robert T. Mason and David Crews, "Female Mimicry in Garter Snakes," *Nature* 316, no. 6023 (1985): 59–60.

10. K. E. Levan, T. Y. Fedina, and S. M. Lewis, "Testing Multiple Hypotheses for the Maintenance of Male Homosexual Copulatory Behaviour in Flour Beetles," *Journal of Evolutionary Biology* 22, no. 1 (2009): 60–70.

11. Ruth E. Buskirk, Cliff Frohlich, and Kenneth G. Ross, "The Natural Selection of Sexual Cannibalism," *American Naturalist* 123 no. 5 (1984): 612–625.

12. Jonathan P. Lelito, and William D. Brown, "Mate Attraction by Females in a Sexually Cannibalistic Praying Mantis," *Behavioral Ecology and Sociobiology* 63, no. 2 (2008): 313–320.

13. Murray P. Fea, Margaret C. Stanley, and Gregory I. Holwell, "Fatal Attraction: Sexually Cannibalistic Invaders Attract Naive Native Mantids," *Biology Letters* 9, no. 6 (2013), http://rsbl.royalsocietypublishing.org/content/roybiolett/9/6/20130746.full .pdf.

14. L. W. Simmons and G. A. Parker, "Nuptial Feeding in Insects: Mating Effort versus Paternal Investment," *Ethology* 81, no. 4 (1989): 332–343.

15. T. Royama, "A Re-interpretation of Courtship Feeding," *Bird Study* 13, no. 2 (1966): 116–129.

16. F. M. Hunter and L. S. Davis, "Female Adélie Penguins Acquire Nest Material from Extrapair Males after Engaging in Extrapair Copulations," *The Auk* 115, no. 2 (1998): 526–528.

17. F. S. Hunter, "Even Penguins Prostitute in a Recession," http://www.styleforum.net/t /90877/even-penguins-prostitute-in-a-recession.

18. Keith Chen, Venkat Lakshminarayanan, and Laurie Santos, "The Dvolution of Our Preferences: Evidence from Capuchin Monkey Trading Behavior," Cowles Foundation Discussion Paper, no. 1524 (2005).

19. Daniel J. Kruger, "Young Adults Attempt Exchanges in Reproductively Relevant Currencies," *Evolutionary Psychology* 6, no. 1 (2008): 204–212.

20. Piotr Tryjanowski and Martin Hromada, "Do Males of the Great Grey Shrike, *Lanius excubitor*, Trade Food for Extrapair Copulations?" *Animal Behaviour* 69, no. 3 (2005): 529–533.

21. Darryl T. Gwynne, "Courtship Feeding in Katydids (Orthoptera: Tettigoniidae): Investment in Offspring or in Obtaining Fertilizations?" *American Naturalist* 128, no. 3 (1986): 342–352.

22. Darryl T. Gwynne, "Mate Selection by Female Katydids (Orthoptera: Tettigoniidae, *Conocephalus nigropleurum*)," *Animal Behaviour* 30, no. 3 (1982): 734–738.

23. R. Robin Baker and Mark A. Bellis, "Human Sperm Competition: Ejaculate Adjustment by Males and the Function of Masturbation," *Animal Behaviour* 46, no. 5 (1993): 861–885.

24. Anne E. Russon, Carel P. van Schaik, P. Kuncoro, A. Ferisa, D. P. Handayani, and M. A. Van Noordwijk, "Innovation and Intelligence in Orangutans," in *Orangutans: Geographic Variation in Behavioral Ecology and Conservation*, ed. Serge A. Wich, S. Suci Utami Atmoko, Tatang Mitra Setia, and Carel P. van Schaik (New York: Oxford University Press, 2009), 279–298.

25. A. De Vos, P. Brokx, and V. Geist, "A Review of Social Behavior of the North American Cervids during the Reproductive Period," *American Midland Naturalist* 77, no. 2 (1967): 390–417.

26. Petra A. Mertens, "Reproductive and Sexual Behavioral Problems in Dogs," *Theriogenology* 66, no. 3 (2006): 606–609; Sue M. McDonnell, M. Henry, and F. Bristol, "Spontaneous Erection and Masturbation in Equids," *Journal of Reproduction and Fertility* 44 (Suppl.), (1991): 664–665.

27. Katherine A. Houpt and Gwendolyn Wollney, "Frequency of Masturbation and Time Budgets of Dairy Bulls Used for Semen Production," *Applied Animal Behaviour Science* 24, no. 3 (1989): 217–225; F. B. M. De Waal, "Bonobo Sex and Society," *Scientific American* 272, no. 3 (1995): 82–88.

28. Baker and Bellis, "Human Sperm Competition," 861–885.

29. Alexander H. Harcourt, Paul H. Harvey, Susan G. Larson, and R. V. Short, "Testis Weight, Body Weight and Breeding System in Primates," *Nature* 293, no. 5827 (1981): 55–57.

30. Ruth Thomsen and Joseph Soltis, "Male Masturbation in Free-ranging Japanese Macaques," *International Journal of Primatology* 25, no. 5 (2004): 1033–1041.

31. Jane M. Waterman, "The Adaptive Function of Masturbation in a Promiscuous African Ground Squirrel," *PloS one* 5, no. 9 (2010): e13060.

32. James Valsa, Kalanghot Padmanabhan Skandhan, Prabhakar Gusani, Pulikkal Sahab Khan, Skandhan Amith, and Meenaxi Gondalia, "Effect of Daily Ejaculation on Semen Quality and Calcium and Magnesium in Semen," *Revista Internacional de Andrología* 11, no. 3 (2013).

5. DO ANIMALS FALL IN LOVE?

1. Jeffrey M. Black and Mark Hulme, *Partnerships in Birds: The Study of Monogamy: The Study of Monogamy* (Oxford: Oxford University Press, 1996).

2. Ulrich H. Reichard and Christophe Boesch, *Monogamy: Mating Strategies and Partnerships in Birds, Humans and Other Mammals* (Cambridge, UK: Cambridge University Press, 2003); Virginia Morell, "A New look at Monogamy," *Science* 281, no. 5385 (1998): 1982.

3. Larry J. Young and Zuoxin Wang, "The Neurobiology of Pair Bonding," *Nature Neuroscience* 7, no. 10 (2004): 1048–1054.

4. C. Nathan De Wall, Omri Gillath, Sarah D. Pressman, Lora L. Black, Jennifer A. Bartz, Jackob Moskovitz, and Dean A. Stetler, "When the Love Hormone Leads to Violence: Oxytocin Increases Intimate Partner Violence Inclinations among High Trait Aggressive People," *Social Psychological and Personality Science* no. 5 (2014): 691–697.

5. Young and Wang, "The Neurobiology of Pair Bonding," 1048–1054.

6. Melvyn S. Soloff, Maria Alexandrova, and Martha J. Fernstrom, "Oxytocin Receptors: Triggers for Parturition and Lactation?" *Science* 204, no. 4399 (1979): 1313–1315.

7. Oliver J. Bosch, Simone L. Meddle, Daniela I. Beiderbeck, Alison J. Douglas, and Inga D. Neumann, "Brain Oxytocin Correlates with Maternal Aggression: Link to Anxiety," *Journal of Neuroscience* 25, no. 29 (2005): 6807–6815.

8. Carsten K. W. De Dreu, Lindred L. Greer, Michel J. J. Handgraaf, Shaul Shalvi, Gerben A. Van Kleef, Matthijs Baas, Femke S. Ten Velden, Eric Van Dijk, and Sander W. W. Feith, "The Neuropeptide Oxytocin Regulates Parochial Altruism in Intergroup Conflict among Humans," *Science* 328, no. 5984 (2010): 1408–1411.

9. Jennifer Bartz, Daphne Simeon, Holly Hamilton, Suah Kim, Sarah Crystal, Ashley Braun, Victor Vicens, and Eric Hollander, "Oxytocin can Hinder Trust and Cooperation in Borderline Personality Disorder," *Social Cognitive and Affective Neuroscience* 6 no. 5 (2011): nsq085.

10. Jennifer A. Bartz, Jamil Zaki, Niall Bolger, and Kevin N. Ochsner, "Social Effects of Oxytocin in Humans: Context and Person Matter," *Trends in Cognitive Sciences* 15, no. 7 (2011): 301–309.

11. Dirk Scheele, Nadine Striepens, Onur Güntürkün, Sandra Deutschländer, Wolfgang Maier, Keith M. Kendrick, and René Hurlemann, "Oxytocin Modulates Social Distance between Males and Females," *Journal of Neuroscience* 32, no. 46 (2012): 16074–16079.

12. Jessie R. Williams, Thomas R. Insel, Carroll R. Harbaugh, and C. Sue Carter, "Oxytocin Administered Centrally Facilitates Formation of a Partner Preference in Female Prairie Voles (Microtus ochrogaster)," *Journal of Neuroendocrinology* 6, no. 3 (1994): 247–250.

13. Beate Ditzen, Marcel Schaer, Barbara Gabriel, Guy Bodenmann, Ulrike Ehlert, and Markus Heinrichs, "Intranasal Oxytocin Increases Positive Communication and Reduces Cortisol Levels During Couple Conflict," *Biological Psychiatry* 65, no. 9 (2009): 728–731.

14. C. Nathan De Wall et al., "When the Love Hormone Leads to Violence."

15. Olga A. Wudarczyk, Brian D. Earp, Adam Guastella, and Julian Savulescu, "Could Intranasal Oxytocin Be Used to Enhance Relationships? Research Imperatives, Clinical Policy, and Ethical Considerations," *Current Opinion in Psychiatry* 26, no. 5 (2013): 474.

16. Charles R. Brown, Mary Bomberger Brown, and Martin L. Shaffer, "Food-sharing Signals Among Socially Foraging Cliff Swallows," *Animal Behaviour* 42, no. 4 (1991): 551–564; John T. Emlen, "Territory, Nest Building, and Pair Formation in the Cliff Swallow," *The Auk* 71, no. 1 (1954): 16–35; Charles R. Brown, *Swallow Summer* (Lincoln: University of Nebraska Press, 1998).

17. Richard H. Wagner, "The Pursuit of Extra-pair Copulations by Female Birds: A New Hypothesis of Colony Formation," *Journal of Theoretical Biology* 163, no. 3 (1993): 333–346.

18. Douglas E. Gladstone, "Promiscuity in Monogamous Colonial Birds," *American Naturalist* 114, no. 4 (1979): 545–557.

19. Joan Roughgarden, *Evolution's Rainbow: Diversity, Gender, and Sexuality in Nature and People* (Berkeley: University of California Press, 2009).

20. Charles R. Brown and Mary Bomberger Brown, "Genetic Evidence of Multiple Parentage in Broods of Cliff Swallows," *Behavioral Ecology and Sociobiology* 23, no. 6 (1988): 379–387.

21. Ibid.

22. Charles R. Brown and Mary Bomberger Brown, "Behavioural Dynamics of Intraspecific Brood Parasitism in Colonial Cliff Swallows," *Animal Behaviour* 37, no. 5 (1989): 777–796.

23. Roughgarden, *Evolution's Rainbow*.

24. Wagner, "The Pursuit of Extra-pair Copulations by Female Birds," 333–346.

25. Tim Clutton-Brock, "Breeding Together: Kin Selection and Mutualism in Cooperative Vertebrates," *Science* 296, no. 5565 (2002): 69–72.

26. Charles R. Brown and Mary Bomberger Brown, *Coloniality in the Cliff Swallow: The Effect of Group Size on Social Behavior* (Chicago: University of Chicago Press, 1996).

27. Brown and Bomberger Brown, "Genetic Evidence of Multiple Parentage in Broods of Cliff Swallows," 379–387.

28. Robert W. Butler, "Wing Fluttering by Mud-gathering Cliff Swallows: Avoidance of 'Rape' Attempts?" *The Auk* 99, no. 4 (1982): 758–761; Charles Robert Brown, *Swallow Summer*.

29. Brown, *Swallow Summer*.

30. Bruce Bagemihl, *Biological Exuberance: Animal Homosexuality and Natural Diversity* (New York: Macmillan, 1999); Roughgarden, *Evolution's Rainbow*.

31. Alistair D. Stutt and Michael T. Siva-Jothy, "Traumatic Insemination and Sexual Conflict in the Bed Bug Cimex lectularius," *Proceedings of the National Academy of Sciences* 98, no. 10 (2001): 5683–5687.

32. Camilla Ryne, "Homosexual Interactions in Bed Bugs: Alarm Pheromones as Male Recognition Signals," *Animal Behaviour* 78, no. 6 (2009): 1471–1475.

33. E. B. Hale, "Visual Stimuli and Reproductive Behavior in Bulls," *Journal of Animal Science* 25, (1966): 36–44.

34. Paul L. Vasey, "Same-sex Sexual Partner Preference in Hormonally and Neurologically Unmanipulated Animals," *Annual Review of Sex Research* 13, no. 1 (2002): 141–179.

35. Bruce Bagemihl, *Biological Exuberance*.

36. Anne Innis Dagg, "Homosexual Behaviour and Female-male Mounting in Mammals—A First Survey," *Mammal Review* 14, no. 4 (1984): 155–185.

37. Srđan Randić, Richard C. Connor, William B. Sherwin, and Michael Krützen, "A Novel Mammalian Social Structure in Indo-Pacific Bottlenose Dolphins (Tursiops sp.): Complex Male Alliances in an Open Social Network," *Proceedings of the Royal Society of London B: Biological Sciences* 282, no. 1813 (2012): rspb20120264.

38. Erin M. Scott, Janet Mann, Jana J. Watson-Capps, Brooke L. Sargeant, and Richard C. Connor, "Aggression in Bottlenose Dolphins: Evidence for Sexual Coercion, Male-Male Competition, and Female Tolerance through Analysis of Tooth-rake Marks and Behaviour," *Behaviour* 142, no. 1 (2005): 21–44.

39. Charles E. Roselli and Fred Stormshak, "The Neurobiology of Sexual Partner Preferences in Rams," *Hormones and Behavior* 55, no. 5 (2009): 611–620.

40. J. A. Resko, A. Perkins, C. E. Roselli, J. N. Stellflug, and F. K. Stormshak, "Sexual Behaviour of Rams: Male Orientation and Its Endocrine Correlates," *Journal of Reproduction and Fertility*. Supplement 54, (1999): 259.

41. Bagemihl, *Biological Exuberance*.

42. Lindsay C. Young, Brenda J. Zaun, and Eric A. VanderWerf, "Successful Same-Sex Pairing in Laysan Albatross," *Biology Letters* 4, no. 4 (2008): 323–325.

43. Justin Richardson and Peter Parnell, *And Tango Makes Three* (New York: Simon and Schuster, 2015).

44. L. Wayne Braithwaite, "Ecological Studies of the Black Swan III. Behaviour and Social Organisation," *Wildlife Research* 8, no. 1 (1981): 135–146.

45. Peter Bogucki, *The Origins of Human Society* (Oxford: Blackwell Publishers, 1999).

46. Katherine Ralls, "Mammals in which Females Are Larger Than Males," *Quarterly Review of Biology* 51, no. 2 (1976): 245–276.

47. L. David Mech, "Alpha Status, Dominance, and Division of Labor in Wolf Packs," *Canadian Journal of Zoology* 77, no. 8 (1999): 1196–1203.

48. Paul L. Vasey, David S. Pocock, and Doug P. VanderLaan, "Kin Selection and Male Androphilia in Samoan Fa'afafine," *Evolution and Human Behavior* 28, no. 3 (2007): 159–167.

49. Andrea Camperio-Ciani, Francesca Corna, and Claudio Capiluppi, "Evidence for Maternally Inherited Factors Favouring Male Homosexuality and Promoting Female Fecundity," *Proceedings of the Royal Society of London* B: Biological Sciences 271, no. 1554 (2004): 2217–2221; Qazi Rahman, Anthony Collins, Martine Morrison, Jennifer Claire Orrells, Khatija Cadinouche, Sherene Greenfield, and Sabina Begum, "Maternal Inheritance and Familial Fecundity Factors in Male Homosexuality," *Archives of Sexual Behavior* 37, no. 6 (2008): 962–969; Francesca Iemmola and Andrea Camperio Ciani, "New Evidence of Genetic Factors Influencing Sexual Orientation in Men: Female Fecundity Increase in the Maternal Line," *Archives of Sexual Behavior* 38, no. 3 (2009): 393–399; Andrea Camperio Ciani, Francesca Iemmola, and Stan R. Blecher, "Genetic Factors Increase Fecundity in Female Maternal Relatives of Bisexual Men as in Homosexuals," *Journal of Sexual Medicine* 6, no. 2 (2009): 449–455; Doug P. VanderLaan and Paul L. Vasey, "Male Sexual Orientation in Independent Samoa: Evidence for Fraternal Birth Order and Maternal Fecundity Effects," *Archives of Sexual Behavior* 40, no. 3 (2011): 495–503.

50. J. A. Resko, A. Perkins, C. E. Roselli, J. N. Stellflug, and F. K. Stormshak, "Sexual Behaviour of Rams, 259.

51. Stella Hu, Angela M. L. Pattatucci, Chavis Patterson, Lin Li, David W. Fulker, Stacey S. Cherny, Leonid Kruglyak, and Dean H. Hamer, "Linkage between Sexual Orientation and Chromosome Xq28 in Males but Not in Females," *Nature Genetics* 11, no. 3 (1995): 248–256.

52. T. Bereczkei, P. Gyuris, P. Koves, and L. Bernath, "Homogamy, Genetic Similarity, and Imprinting; Parental Influence on Mate Choice Preferences," *Personality and Individual Differences* 33, no. 5 (2002): 677–690.

53. A. C. Little, I. S. Penton-Voak, D. M. Burt, and D. I. Perrett, "Investigating an Imprinting-like Phenomenon in Humans: Partners and Opposite-sex Parents Have Similar Hair and Eye Colour," *Evolution and Human Behavior* 24, no. 1 (2003): 43–51.

54. Agnieszka Wiszewska, Boguslaw Pawlowski, and Lynda G. Boothroyd, "Father–Daughter Relationship as a Moderator of Sexual Imprinting: A Facialmetric Study," *Evolution and Human Behavior* 28, no. 4 (2007): 248–252; Tamas Bereczkei, Petra Gyuris, and Glenn E. Weisfeld, "Sexual Imprinting in Human Mate Choice," *Proceedings of the Royal Society of London*, Series B: Biological Sciences 271, no. 1544 (2004): 1129–1134; Glenn D. Wilson and Paul T. Barrett, "Parental Characteristics and Partner Choice: Some Evidence for Oedipal Imprinting," *Journal of Biosocial Science* 19, no. 2 (1987): 157–161.

55. Edward Westermarck, *The History of Human Marriage* (New York: Macmillan, 1921).

56. Joseph Shepher, "Mate Selection among Second Generation Kibbutz Adolescents and Adults: Incest Avoidance and Negative Imprinting," *Archives of Sexual Behavior* 1, no. 4 (1971): 293–307.

57. Ibid.

58. Melford E. Spiro, *Children of the Kibbutz* (Cambridge, Mass.: Harvard University Press, 1958).

59. Eran Shor and Dalit Simchai, "Incest Avoidance, the Incest Taboo, and Social Cohesion: Revisiting Westermarck and the Case of the Israeli Kibbutzim," *American Journal of Sociology* 114, no. 6 (2009): 1803–1842.

60. David F. Aberle, Urie Bronfenbrenner, Eckhard H. Hess, Daniel R. Miller, David M. Schneider, and James N. Spuhler, "The Incest Taboo and the Mating Patterns of Animals," *American Anthropologist* 65, no. 2 (1963): 253–265.

61. Jane Goodall, *The Chimpanzees of Gombe: Patterns of Behavior* (Cambridge, Mass.: Belknap Press of Harvard University Press, 1986).

62. Lewis Thomas, *The Lives of a Cell: Notes of a Biology Watcher* (New York: Viking, 1978).

63. Suma Jacob, Martha K. McClintock, Bethanne Zelano, and Carole Ober, "Paternally Inherited HLA Alleles are Associated with Women's Choice of Male Odor," *Nature Genetics* 30, no. 2 (2002): 175–179.

64. Trese Leinders-Zufall, Peter Brennan, Patricia Widmayer, Andrea Maul-Pavicic, Martina Jäger, Xiao-Hong Li, Heinz Breer, Frank Zufall, and Thomas Boehm, "MHC Class I Peptides as Chemosensory Signals in the Vomeronasal Organ," *Science* 306, no. 5698 (2004): 1033–1037.

65. T. Tregenza and N. Wedell, "Genetic Compatibility, Mate Choice and Patterns of Parentage: Invited Review," *Molecular Ecology* 9, no. 8 (2000): 1013–1027.

66. Claus Wedekind, Thomas Seebeck, Florence Bettens, and Alexander J. Paepke, "MHC-dependent Mate Preferences in Humans," *Proceedings of the Royal Society of London*, Series B: Biological Sciences 260, no. 1359 (1995): 245–249.

67. Jacob et al., "Paternally Inherited HLA Alleles are Associated with Women's Choice of Male Odor," 175–179.

68. Wedekind et al., "MHC-dependent Mate Preferences in Humans," 245–249.

6. THE AGONY OF GRIEF

1. Jane Goodall, *Through a Window: My Thirty Years with the Chimpanzees of Gombe* (Boston: Houghton Mifflin Harcourt, 2010).

2. Barbara J. King, *How Animals Grieve* (Chicago: University of Chicago Press, 2013).

3. Diana Reiss and Lori Marino, "Mirror Self-recognition in the Bottlenose Dolphin: A Case of Cognitive Convergence," *Proceedings of the National Academy of Sciences* 98, no. 10 (2001): 5937–5942.

4. Rowan Hooper, "Dolphins Appear to Grieve in Different Ways," *New Scientist* 211, no. 2828 (2011): 10.

5. Marc Bekoff, "Grief in Animals: It's Arrogant to Think We're the Only Animals Who Mourn," *Psychology Today*, https://www.psychologytoday.com/blog/animal-emotions/200910/grief-in-animals-its-arrogant-think-were-the-only-animals-who-mourn.

6. Francesco Mazzini, Simon W. Townsend, Zsófia Virányi, and Friederike Range, "Wolf Howling is Mediated by Relationship Quality Rather than Underlying Emotional Stress," *Current Biology* 23, no. 17 (2013): 1677–1680.

7. Rennie Bere, *The African Elephant* (West Sussex, UK: Littlehampton Book Services, 1966).

8. Joyce Poole, *Coming of Age with Elephants: A Memoir* (New York: Hyperion, 1996).

9. Cavan Sieczkowski, "Baby Elephant Cries for 5 Hours after Mom Attacks, Rejects Him," *The Huffington Post*, http://www.huffingtonpost.com/2013/09/13/baby-elephant-cries_n_3920685.html.

10. Sutapa Mukerjee, "Elephant Who Could Not Forget Dies of Broken Heart," *The Guardian*, http://www.theguardian.com/world/1999/may/07/5.

11. Reuters, "Elephant Accidentally Kills Missouri Zoo Keeper," *The Telegraph*, http://www.telegraph.co.uk/news/earth/wildlife/10374525/Elephant-accidentally-kills-Missouri-zoo-keeper.html.

12. Natalie Angier, "About Death, Just Like Us or Pretty Much Unaware?," *New York Times*, http://www.nytimes.com/2008/09/02/science/02angi.html.

13. Carrie Packwood Freeman, Marc Bekoff, and Sarah M. Bexell, "Giving Voice to the 'Voiceless:' Incorporating Nonhuman Animal Perspectives as Journalistic Sources," *Journalism Studies* 12, no. 5 (2011): 590–607.

14. Ivan Petrovich Pavlov, *Conditioned Reflexes: An Investigation of the Physiological Activity of the Cerebral Cortex* (Oxford: Oxford University Press, 1927), 397.

15. Marc Bekoff and Jane Goodall, *Minding Animals: Awareness, Emotions, and Heart* (New York: Oxford University Press, 2002).

16. Robert W. Mitchell, Nicholas S. Thompson, and H. Lyn Miles, *Anthropomorphism, Anecdotes, and Animals* (Albany: State University of New York Press, 1997).

17. Vicki Hamilton, Karen Evans, Ben Raymond, and Mark A. Hindell, "Environmental Influences on Tooth Growth in Sperm Whales from Southern Australia," *Journal of Experimental Marine Biology and Ecology* 446 (2013): 236–244.

18. Mary-Frances O'Connor, Michael R. Irwin, and David K. Wellisch, "When Grief Heats Up: Pro-inflammatory Cytokines Predict Regional Brain Activation," *Neuroimage* 47, no. 3 (2009): 891–896.

19. C. Sue Carter, "Neuroendocrine Perspectives on Social Attachment and Love," *Psychoneuroendocrinology* 23, no. 8 (1998): 779–818; Jaak Panksepp, "Oxytocin Effects on Emotional Processes: Separation Distress, Social Bonding, and Relationships to Psychiatric Disorders," *Annals of the New York Academy of Sciences* 652, no. 1 (1992): 243–252.

20. Miranda M. Lim and Larry J. Young, "Neuropeptidergic Regulation of Affiliative Behavior and Social Bonding in Animals," *Hormones and Behavior* 50, no. 4 (2006): 506–517.

21. James R. Averill, *Emotions in Personality and Psychopathology* (New York: Plenum, 1979), 337–368.

22. John Archer, "Grief from an Evolutionary Perspective," in *Handbook of Bereavement Research: Consequences, Coping, and Care*, ed. Margaret S. Stroebe (Washington, D.C: American Psychological Association, 2001): 263–83.

23. Anne L. Engh, Jacinta C. Beehner, Thore J. Bergman, Patricia L. Whitten, Rebekah R. Hoffmeier, Robert M. Seyfarth, and Dorothy L. Cheney, "Behavioural and Hormonal Responses to Predation in Female Chacma Baboons (Papio hamadryas ursinus)," *Proceedings of the Royal Society B: Biological Sciences* 273, no. 1587 (2006): 707–712.

24. University of Pennsylvania, "Baboons in Mourning Seek Comfort among Friends," *ScienceDaily* (2006), http://www.sciencedaily.com/releases/2006/01/060130154735.htm.

25. Joyce Poole, *Coming of Age with Elephants: A Memoir* (New York: Hyperion, 1996).

26. Martin Meredith, *Elephant Destiny: Biography of an Endangered Species in Africa* (New York: Public Affairs, 2004), available at http://onlinelibrary.wiley.com/doi/10.1111/j.1526-4629.2004.tb00111.x/pdf.

27. Karen McComb, Lucy Baker, and Cynthia Moss, "African Elephants Show High Levels of Interest in the Skulls and Ivory of Their Own Species," *Biology Letters* 2, no. 1 (2006): 26–28.

28. Michael Parker Pearson, *The Archaeology of Death and Burial* (Phoenix Mill, UK: Sutton, 1999).

29. William Rendu, Cédric Beauval, Isabelle Crevecoeur, Priscilla Bayle, Antoine Balzeau, Thierry Bismuth, Laurence Bourguignon, Géraldine Delfour, Jean-Philippe Faivre, and François Lacrampe-Cuyaubère, "Evidence Supporting an Intentional Neandertal Burial at La Chapelle-aux-Saints," *Proceedings of the National Academy of Sciences* 111, no. 1 (2014): 81–86.

30. Paul Pettitt, *The Palaeolithic Origins of Human Burial* (London: Routledge, 2013).

31. Dian Fossey, *Gorillas in the Mist* (Boston: Houghton Mifflin Harcourt, 2000).

32. William Mullen, "One by One, Gorillas Pay Their Last Respects," *The Chicaco Tribune*, http://articles.chicagotribune.com/2004–12–08/news/0412080315_1_babs-gorilla -brookfield-zoo.

33. T. L. Iglesias, R. McElreath, and G. L. Patricelli, "Western Scrub-Jay Funerals: Cacophonous Aggregations in Response to Dead Conspecifics," *Animal Behaviour* 84, no. 5 (2012): 1103–1111.

34. Marc Bekoff, "Grieving Animals: Saying Goodbye to Friends and Family," *Psychology Today*, https://www.psychologytoday.com/blog/animal-emotions/201207/grieving-animals -saying-goodbye-friends-and-family.

35. Marc Bekoff, "Are You Feeling What I'm Feeling?" *New Scientist* 194, no. 2605 (2007): 42–47.

36. Barbara J. King, "When Animals Mourn," *Scientific American* 309, no. 1 (2013): 62–67.

37. Konrad Lorenz, Michael Martys, and Angelika Tipler, *Here Am I: Where Are You?: The Behavior of the Greylag Goose* (San Diego: Harcourt Brace, 1991).

38. Jerry Jacob, "Grief-stricken Goose Mourns Loss at Dollar General," *Ky3*, http://articles .ky3.com/2012–05–29/canada-geese_31891220.

39. Carol Buckley, *Tarra & Bella: The Elephant and Dog Who Became Best Friends* (London: Penguin, 2009).

40. Francine Patterson, *Koko's Kitten* (New York: Scholastic, 1995).

7. JEALOUS BEASTS

1. Esther Herrmann, Josep Call, María Victoria Hernández-Lloreda, Brian Hare, and Michael Tomasello, "Humans Have Evolved Specialized Skills of Social Cognition: The Cultural Intelligence Hypothesis," *Science* 317, no. 5843 (2007): 1360–1366; Anne E. Russon, "Naturalistic Approaches to Orangutan Intelligence and the Question of Enculturation," *International Journal of Comparative Psychology* 12, no. 4 (1999).

2. http://www.elpasozoo.org/Press-Release-29.php, accessed July 3, 2013.

3. "Killed By Gorilla," *Alexandria Gazette*, March 10, 1902).

4. J. Wong and Nico K. Michiels, "Control of Social Monogamy through Aggression in a Hermaphroditic Shrimp," *Front Zool* 8, no. 30 (2011): 30–37.

5. Marta M. Rufino and David A. Jones, "Binary Individual Recognition in Lysmata Debelius (Decapoda: Hippolytidae) under Laboratory Conditions," *Journal of Crustacean Biology* 21, no. 2 (2001): 388–392.

6. Sally P. Mendoza and William A. Mason, "Parental Division of Labour and Differentiation of Attachments in a Monogamous Primate (*Callicebus moloch*)," *Animal Behaviour* 34, no. 5 (1986): 1336–1347.

7. D. D. Cubicciotti III and W. A. Mason, "Comparative Studies of Social Behavior in Callicebus and Saimiri: Heterosexual Jealousy Behavior," *Behavioral Ecology and Sociobiology* 3, no. 3 (1978): 311–322.

8. Dorothy L. Cheney and Robert M. Seyfarth, *Baboon Metaphysics: The Evolution of a Social Mind* (Chicago: University of Chicago Press, 2008).

9. Susan C. Alberts, Jeanne Altmann, and Michael L. Wilson, "Mate Guarding Constrains Foraging Activity of Male Baboons," *Animal Behaviour* 51, no. 6 (1996): 1269–1277.

10. Ryne A. Palombit and Julia Fischer Cheney, "Male Infanticide and Defense of Infants in Chacma Baboons," in *Infanticide by Males and Its Implications*, ed. Carel van Schaik and Charles Janson (Cambridge: Cambridge University Press, 2000): 123–152.

11. David P. Watts, "Infanticide in Mountain Gorillas: New Cases and a Reconsideration of the Evidence," *Ethology* 81, no. 1 (1989): 1–18.

12. "Third Tragedy Hits London Zoo's Gorillas: First Two Males Die . . . Now Baby Tiny is Crushed by Jealous Silverback," *The Daily Mail*, 14 May 2011.

13. Barry S. Hewlett, *Father-Child Relations: Cultural and Biosocial Contexts* (New Brunswick, N.J.: Transaction Publishers, 1992).

14. David M. Buss, Randy J. Larsen, Drew Westen, and Jennifer Semmelroth, "Sex Differences in Jealousy: Evolution, Physiology, and Psychology," *Psychological Science* 3, no. 4 (1992): 251–255.

15. Barbara Smuts, "The Evolutionary Origins of Patriarchy," *Human Nature* 6, no. 1 (1995): 1–32.

16. A. Cooper and E. L. Smith. "Homicide in the U.S. Known to Law Enforcement, 2011" (Washington D.C.: Bureau of Justice Statistics, 2013).

17. David M. Buss, *The Murderer Next Door: Why the Mind is Designed to Kill* (New York: Penguin, 2006).

18. Susan Brownmiller, *Against Our Will: Men, Women and Rape* (New York: Open Road Media, 2013).

19. James K. Rilling, James T. Winslow, and Clinton D. Kilts, "The Neural Correlates of Mate Competition in Dominant Male Rhesus Macaques," *Biological Psychiatry* 56, no. 5 (2004): 364–375.

20. Karen L. Bales, William A. Mason, Ciprian Catana, Simon R. Cherry, and Sally P. Mendoza, "Neural Correlates of Pair-bonding in a Monogamous Primate," *Brain Research* 1184 (2007): 245–253.

21. Hidehiko Takahashi, Masato Matsuura, Noriaki Yahata, Michihiko Koeda, Tetsuya Suhara, and Yoshiro Okubo, "Men and Women Show Distinct Brain Activations during Imagery of Sexual and Emotional Infidelity," *NeuroImage* 32, no. 3 (2006): 1299–1307.

22. Heidi Greiling and David M. Buss, "Women's Sexual Strategies: The Hidden Dimension of Extra-pair Mating," *Personality and Individual Differences* 28, no. 5 (2000): 929–963.

23. Bram P. Buunk and Pieternel Dijkstra, "Men, Women, and Infidelity: Sex Differences in Extradyadic Sex and Jealousy," in *The State of Affairs: Explorations in Infidelity and Commitment*, ed. J. Duncombe, K. Harrison, G. Allan, and D. Marsden (New York: Routledge, 2004), 103–120.

24. Steven W. Gangestad, Randy Thornhill, and Christine E. Garver, "Changes in Women's Sexual Interests and Their Partner's Mate–retention Tactics across the Menstrual Cycle:

Evidence for Shifting Conflicts of Interest," *Proceedings of the Royal Society of London,* Series B: Biological Sciences 269, no. 1494 (2002): 975–982.

25. Nancy Burley, "The Evolution of Concealed Ovulation," *American Naturalist* 114, no. 6 (1979): 835–858.

26. Brad J. Sagarin, D. Vaughn Becker, Rosanna E. Guadagno, Wayne W. Wilkinson, and Lionel D. Nicastle, "A Reproductive Threat-based Model of Evolved Sex Differences in Jealousy," *Evolutionary Psychology* 10, no. 3 (2012): 487.

27. Christopher J. Carpenter, "Meta-analyses of Sex Differences in Responses to Sexual versus Emotional Infidelity in Men and Women Are More Similar than Different," *Psychology of Women Quarterly* 36, no. 1 (2012): 25–37.

28. Brian C. Bertram, "The Social System of Lions," *Scientific American* 232, no. 5 (1975).

8. DARKER STILL

1. John T. Emlen, "Social Behavior in Nesting Cliff Swallows," *The Condor* 54, no. 4 (1952): 177–199.

2. John L. Hoogland and Paul W. Sherman, "Advantages and Disadvantages of Bank Swallow (Riparia riparia) Coloniality," *Ecological Monographs* 46, no. 1 (1976): 33–58.

3. Charles R. Brown and Mary Bomberger Brown, "Selection of High-quality Host Nests by Parasitic Cliff Swallows," *Animal Behaviour* 41, no. 3 (1991): 457–465.

4. Angela D. Bryan, Gregory D. Webster, and Amanda L. Mahaffey, "The Big, the Rich, and the Powerful: Physical, Financial, and Social Dimensions of Dominance in Mating and Attraction," *Personality and Social Psychology Bulletin* 37, no. 3 (2011): 365–382.

5. Nadia R. Bardack and Francis T. McAndrew, "The Influence of Physical Attractiveness and Manner of Dress on Success in a Simulated Personnel Decision," *Journal of Social Psychology* 125, no. 6 (1985): 777–778.

6. Matthew Mulford, John Orbell, Catherine Shatto, and Jean Stockard, "Physical Attractiveness, Opportunity, and Success in Everyday Exchange 1," *American Journal of Sociology* 103, no. 6 (1998): 1565–1592.

7. Chris J. Boyatzis, Peggy Baloff, and Cheri Durieux, "Effects of Perceived Attractiveness and Academic Success on Early Adolescent peer Popularity," *Journal of Genetic Psychology* 159, no. 3 (1998): 337–344.

8. Larry L. Wolf and F. Reed Hainsworth, "Time and Energy Budgets of Territorial Hummingbirds," *Ecology* 52, no. 6 (1971): 980–988.

9. John Byers, Eileen Hebets, and Jeffrey Podos, "Female Mate Choice Based upon Male Motor Performance," *Animal Behaviour* 79, no. 4 (2010): 771–778.

10. Jerram L. Brown, "The Evolution of Diversity in Avian Territorial Systems," *Wilson Bulletin* 76, no. 2 (1964): 160–169.

11. Celia Haigh Holm, "Breeding Sex Ratios, Territoriality, and Reproductive Success in the Red-winged Blackbird (Agelaius phoeniceus)," *Ecology* 54, no. 2 (1973): 356–365.

12. J. Roughgarden, *Evolution's Rainbow: Diversity, Gender, and Sexuality in Nature and People* (Berkeley: University of California Press, 2009).

13. William F. Wood, Miranda N. Terwilliger, and Jeffrey P. Copeland, "Volatile Compounds from Anal Glands of the Wolverine, Gulo gulo," *Journal of Chemical Ecology* 31, no. 9 (2005): 2111–2117.

14. Hiroshi Yamada, Agnieszka Tymula, Kenway Louie, and Paul W. Glimcher, "Thirst-dependent Risk Preferences in Monkeys Identify a Primitive Form of Wealth," *Proceedings of the National Academy of Sciences* 110, no. 39 (2013): 15788–15793.

15. Charles A. Holt and Susan K. Laury, "Risk Aversion and Incentive Effects," *American Economic Review* 92, no. 5 (2002): 1644–1655.

16. Kristy van Marle, Justine Aw, Koleen McCrink, and Laurie R. Santos, "How Capuchin Monkeys (Cebus apella) Quantify Objects and Substances," *Journal of Comparative Psychology* 120, no. 4 (2006): 416.

17. M. Keith Chen, Venkat Lakshminarayanan, and Laurie R. Santos, "How Basic Are Behavioral Biases? Evidence from Capuchin Monkey Trading Behavior," *Journal of Political Economy* 114, no. 3 (2006): 517–537.

18. Venkat Lakshminaryanan, M. Keith Chen, and Laurie R. Santos, "Endowment Effect in Capuchin Monkeys," *Philosophical Transactions of the Royal Society B: Biological Sciences* 363, no. 1511 (2008): 3837–3844.

19. Richard B. D'Eath, Bert J. Tolkamp, Ilias Kyriazakis, and Alistair B. Lawrence, "'Freedom from Hunger' and Preventing Obesity: The Animal Welfare Implications of Reducing Food Quantity or Quality," *Animal Behaviour* 77, no. 2 (2009): 275–288.

20. Amy Luke, Lara R. Dugas, Kara Ebersole, Ramon A Durazo-Arvizu, Guichan Cao, Dale A. Schoeller, Adebowale Adeyemo, William R. Brieger, and Richard S. Cooper, "Energy Expenditure Does Not Predict Weight Change in Either Nigerian or African American Women," *The American Journal of Clinical Nutrition* 89, no. 1 (2009): 169–176.

21. Peter N. Stearns, *Fat History: Bodies and Beauty in the Modern West* (New York: NYU Press, 2002).

22. L. J. Arone, Ronald Mackintosh, Michael Rosenbaum, Rudolph L. Leibel, and Jules Hirsch, "Autonomic Nervous System Activity in Weight Gain and Weight Loss," *American Journal of Physiology-Regulatory, Integrative and Comparative Physiology* 269, no. 1 (1995): R222–R225.

23. Eric Ravussin, Stephen Lillioja, William C. Knowler, Laurent Christin, Daniel Freymond, William G. H. Abbott, Vicky Boyce, Barbara V. Howard, and Clifton Bogardus, "Reduced Rate of Energy Expenditure as a Risk Factor for Body-weight Gain," *New England Journal of Medicine* 318, no. 8 (1988): 467–472.

9. AFRAID OF THE DARK

1. Robert Plutchik, "The Nature of Emotions: Human Emotions Have Deep Evolutionary Roots, a Fact That May Explain Their Complexity and Provide Tools for Clinical Practice," *American Scientist* 89, no. 4 (2001): 344–350.

2. Charles Darwin, *The Expression of the Emotions in Man and Animals* (New York: D. Appleton, 1886), 38.

3. Peter J. Lang, Margaret M. Bradley, and Bruce N. Cuthbert, "Emotion, Attention, and the Startle Reflex," *Psychological Review* 97, no. 3 (1990): 377.

4. Robert C. Eaton, *Neural Mechanisms of Startle Behavior* (New York: Springer Science & Business Media, 2013).

5. Edmund D. Brodie, "Predator-Prey Arms Races: Asymmetrical Selection on Predators and Prey May be Reduced When Prey Are Dangerous," *Bioscience* 49, no. 7 (1999): 557–568.

6. John B. Watson, and Rosalie Rayner, "Conditioned Emotional Reactions," *Journal of Experimental Psychology* 3, no. 1 (1920): 1.

7. Markus Fendt and Michael S. Fanselow, "The Neuroanatomical and Neurochemical Basis of Conditioned Fear," *Neuroscience & Biobehavioral Reviews* 23, no. 5 (1999): 743–760.

8. Michael Davis, "The Role of the Amygdala in Fear and Anxiety," *Annual Review of Neuroscience* 15, no. 1 (1992): 353–375.

9. Ahmad R. Hariri, Susan Y. Bookheimer, and John C. Mazziotta, "Modulating Emotional Responses: Effects of a Neocortical Network on the Limbic System," *Neuroreport* 11, no. 1 (2000): 43–48.

10. Frederick A. King, "Effects of Septal and Amygdaloid Lesions on Emotional Behavior and Conditioned Avoidance Responses in the Rat," *Journal of Nervous and Mental Disease* 126, no. 1 (1958): 57–63.

11. Mark Barad, Po-Wu Gean, and Beat Lutz, "The Role of the Amygdala in the Extinction of Conditioned Fear," *Biological Psychiatry* 60, no. 4 (2006): 322–328.

12. Stefan G. Hofmann, "Cognitive Processes during Fear Acquisition and Extinction in Animals and Humans: Implications for Exposure Therapy of Anxiety Disorders," *Clinical Psychology Review* 28, no. 2 (2008): 199–210.

13. Gregory S. Berns, Jonathan Chappelow, Milos Cekic, Caroline F. Zink, Giuseppe Pagnoni, and Megan E. Martin-Skurski, "Neurobiological Substrates of Dread," *Science* 312, no. 5774 (2006): 754–758.

14. Giles W. Story, Ivaylo Vlaev, Ben Seymour, Joel S. Winston, Ara Darzi, and Raymond J. Dolan, "Dread and the Disvalue of Future Pain," *PLoS Computational Biology* 9, no. 11 (2013): e1003335; Simon Makin, "Waiting for Pain Can Cause More Dread Than Pain Itself," *New Scientist* 220, no. 2945 (2013): 16.

15. Andreas Olsson, Katherine I. Nearing, and Elizabeth A. Phelps, "Learning Fears by Observing Others: The Neural Systems of Social Fear Transmission," *Social Cognitive and Affective Neuroscience* 2, no. 1 (2007): 3–11.

16. Andreas Olsson and Elizabeth A. Phelps, "Social Learning of Fear," *Nature Neuroscience* 10, no. 9 (2007): 1095–1102.

17. Mark J. Boschen, "Reconceptualizing Emetophobia: A Cognitive–Behavioral Formulation and Research Agenda," *Journal of Anxiety Disorders* 21, no. 3 (2007): 407–419.

18. Vincent Paquette, Johanne Lévesque, Boualem Mensour, Jean-Maxime Leroux, Gilles Beaudoin, Pierre Bourgouin, and Mario Beauregard, "'Change the Mind and You Change the Brain': Effects of Cognitive-Behavioral Therapy on the Neural Correlates of Spider Phobia," *Neuroimage* 18, no. 2 (2003): 401–409.

19. Aaron T. Beck, Gary Emery, and Ruth L. Greenberg, *Anxiety Disorders and Phobias: A Cognitive Perspective* (Cambridge, Mass.: Basic Books, 2005); Jan Resnick, "Far-out Philias and Phobias," *Psychotherapy in Australia* 17, no. 4 (2011): 65; and "The Ultimate List of Phobias and Fears," http://www.fearof.net.

20. Karen L. Overall, Arthur E. Dunham, and Diane Frank, "Frequency of Nonspecific Clinical Signs in Dogs with Separation Anxiety, Thunderstorm Phobia, and Noise Phobia, Alone or in Combination," *Journal of the American Veterinary Medical Association* 219, no. 4 (2001): 467–473.

21. David M. Clark, Anke Ehlers, Ann Hackmann, Freda McManus, Melanie Fennell, Nick Grey, Louise Waddington, and Jennifer Wild, "Cognitive Therapy versus Exposure and Applied Relaxation in Social Phobia: A Randomized Controlled Trial," *Journal of Consulting and Clinical Psychology* 74, no. 3 (2006): 568.

22. Arne Öhman, "Face the Beast and Fear the Face: Animal and Social Fears as Prototypes for Evolutionary Analyses of Emotion," *Psychophysiology* 23, no. 2 (1986): 123–145.

23. Douglas W. Morrison, "Lunar Phobia in a Neotropical Fruit Bat, Artibevs jamaicensis (Chiroptera: Phyllostomidae)," *Animal Behaviour* 26, no. 3 (1978): 852–855.

24. Sharon Gursky, "Lunar Philia in a Nocturnal Primate," *International Journal of Primatology* 24, no. 2 (2003): 351–367.

25. Sharon Gursky and K.A.I. Nekaris, *Primate Anti-predator Strategies* (New York: Springer, 2007).

26. Martin E. P. Seligman, "Phobias and Preparedness," *Behavior Therapy* 2, no. 3 (1971): 307–320.

27. Judy S. DeLoache and Vanessa LoBue, "The Narrow Fellow in the Grass: Human Infants Associate Snakes and Fear," *Developmental Science* 12, no. 1 (2009): 201–207.

28. Glenn E. King, "The Attentional Basis for Primate Responses to Snakes," Annual Meeting of the American Society of Primatologists, San Diego, California (June 1997).

29. Michael Cook and Susan Mineka, "Selective Associations in the Observational Conditioning of Fear in Rhesus Monkeys," *Journal of Experimental Psychology: Animal Behavior Processes* 16, no. 4 (1990): 372.

30. Susan Mineka, "A Primate Model of Phobic Fears," in *Theoretical Foundations of Behavior Therapy*, ed. H. Eysenck and I. Martin (New York: Springer, 1987), 81–111.

31. Lynne A. Isbell, "Snakes as Agents of Evolutionary Change in Primate Brains," *Journal of Human Evolution* 51, no. 1 (2006): 1–35.

32. Charles Siebert, "Orphan Elephants," *National Geographic*, September 2011.
33. Gay A. Bradshaw, Allan N. Schore, Janine L. Brown, Joyce H. Poole, and Cynthia J. Moss, "Elephant Breakdown," *Nature* 433, no. 7028 (2005): 807–807; Graeme Shannon, Rob Slotow, Sarah M. Durant, Katito N. Sayialel, Joyce Poole, Cynthia Moss, and Karen McComb, "Effects of Social Disruption in Elephants Persist Decades after Culling," *Frontiers in Zoology* 10, no. 1 (2013): 62.
34. Hope R. Ferdowsian, Debra L. Durham, Charles Kimwele, Godelieve Kranendonk, Emily Otali, Timothy Akugizibwe, J. B. Mulcahy, Lilly Ajarova, and Cassie Meré Johnson, "Signs of Mood and Anxiety Disorders in Chimpanzees," *PLoS One* 6, no. 6 (2011): e19855.
35. Richard G. Lister, "Ethologically-based Animal Models of Anxiety Disorders," *Pharmacology & Therapeutics* 46, no. 3 (1990): 321–340.

10. THE RICHNESS OF ANIMAL COMMUNICATION

1. Jared Diamond, "The Great Leap Forward," *Discover Magazine*, October 1989, 15–23.
2. Constance Holden, "The Origin of Speech," *Science* 303, no. 5662 (2004): 1316–1319; Elizabeth Pennisi, "The First Language?," *Science* 303, no. 5662 (2004): 1319–1320.
3. Barbara Pease and Allan Pease, *The Definitive Book of Body Language* (New York: Bantam, 2004).
4. Donald D. Price, Ronald Dubner, and James W Hu, "Trigeminothalamic Neurons in Nucleus Caudalis Responsive to Tactile, Thermal, and Nociceptive Stimulation of Monkey's Face," *Journal of Neurophysiology* 39, no. 5 (1976): 936–953; Richard A. Depue and Jeannine V. Morrone-Strupinsky, "A Neurobehavioral Model of Affiliative Bonding: Implications for Conceptualizing a Human Trait of Affiliation," *Behavioral and Brain Sciences* 28, no. 3 (2005): 313–349.
5. John R. Krebs and Richard Dawkins, "Animal Signals: Mind-Reading and Manipulation," in *Behavioural Ecology: An Evolutionary Approach*, ed. J. R. Krebs and N. B. Davies (Oxford: Blackwell, 1984), 380–402.
6. Signe Preuschoft, "Primate Faces and Facial Expressions," *Social Research* 67, no. 1 (2000): 245–271.
7. Harry Walter Greene, "Antipredator Mechanisms in Reptiles," *Biology of the Reptilia* 16 (1988): 1–152.
8. Rufus A. Johnstone, "Honest Advertisement of Multiple Qualities Using Multiple Signals," *Journal of Theoretical Biology* 177, no. 1 (1995): 87–94.
9. Claire D. FitzGibbon and John H. Fanshawe, "Stotting in Thomson's Gazelles: An Honest Signal of Condition," *Behavioral Ecology and Sociobiology* 23, no. 2 (1988): 69–74.
10. Ibid.
11. T. M. Caro, "The Functions of Stotting: A Review of the Hypotheses," *Animal Behaviour* 34, no. 3 (1986): 649–662.

12. Oren Hasson, "Pursuit-Deterrent Signals: Communication Between Prey and Predator," *Trends in Ecology & Evolution* 6, no. 10 (1991): 325–329.

13. Klaus Zuberbühler, Ronald Noë, and Robert M. Seyfarth, "Diana Monkey Long-distance Calls: Messages for Conspecifics and Predators," *Animal Behaviour* 53, no. 3 (1997): 589–604.

14. Paul Ekman, Wallace V. Friesen, Maureen O'Sullivan, Anthony Chan, Irene Diacoyanni-Tarlatzis, Karl Heider, Rainer Krause, William Ayhan LeCompte, Tom Pitcairn, and Pio E. Ricci-Bitti, "Universals and Cultural Differences in the Judgments of Facial Expressions of Emotion," *Journal of Personality and Social Psychology* 53, no. 4 (1987): 712.

15. Dale J. Langford, Andrea L. Bailey, Mona Lisa Chanda, Sarah E. Clarke, Tanya E. Drummond, Stephanie Echols, Sarah Glick, Joelle Ingrao, Tammy Klassen-Ross, and Michael L. LaCroix-Fralish, "Coding of Facial Expressions of Pain in the Laboratory Mouse," *Nature Methods* 7, no. 6 (2010): 447–449.

16. Susana G. Sotocinal, Robert E. Sorge, Austin Zaloum, Alexander H. Tuttle, Loren J. Martin, Jeffrey S. Wieskopf, Josiane C. S. Mapplebeck, Peng Wei, Shu Zhan, and Shuren Zhang, "The Rat Grimace Scale: A Partially Automated Method for Quantifying Pain in the Laboratory Rat via Facial Expressions," *Molecular Pain* 7, no. 1 (2011): 55.

17. Amanda C. de C. Williams, "Facial Expression of Pain, Empathy, Evolution, and Social Learning," *Behavioral and Brain Sciences* 25, no. 04 (2002): 475–480.

18. Larry Gray, Lisa W. Miller, Barbara L. Philipp, and Elliott M. Blass, "Breastfeeding Is Analgesic in Healthy Newborns," *Pediatrics* 109, no. 4 (2002): 590–593.

19. Williams, "Facial Expression of Pain, Empathy, Evolution, and Social Learning," 475–480; Thomas Hadjistavropoulos, Kenneth D. Craig, Steve Duck, Annmarie Cano, Liesbet Goubert, Philip L. Jackson, Jeffrey S. Mogil, Pierre Rainville, Michael J. L. Sullivan, and Amanda C. de C. Williams, "A Biopsychosocial Formulation of Pain Communication," *Psychological Bulletin* 137, no. 6 (2011): 910.

20. Elisabeth H. M. Sterck, and Brigitte M. A. Goossens, "The Meaning of "Macaque" Facial Expressions," *Proceedings of the National Academy of Sciences USA* 105, no. (2008): E71; Dario Maestripieri, "Gestural Communication in Macaques: Usage and Meaning of Nonvocal Signals," *Evolution of Communication* 1, no. 2 (1997): 193–222.

21. Mary L. Phillips, Andy W. Young, Carl Senior, Michael Brammer, Chris Andrew, Andrew J. Calder, Edward T. Bullmore, D. I. Perrett, D. Rowland, and S. C. R. Williams, "A Specific Neural Substrate for Perceiving Facial Expressions of Disgust," *Nature* 389, no. 6650 (1997): 495–498.

22. Gisela Kaplan, and Lesley J. Rogers, "Patterns of Gazing in Orangutans (Pongo pygmaeus)," *International Journal of Primatology* 23, no. 3 (2002): 501–526.

23. Marina Davila Ross, Susanne Menzler, and Elke Zimmermann, "Rapid Facial Mimicry in Orangutan Play," *Biology Letters* 4, no. 1 (2008): 27–30.

24. Adam Kendon, Thomas A Sebeok, and Jean Umiker-Sebeok, *Nonverbal Communication, Interaction, and Gesture: Selections from Semiotica* (The Hague: Mouton Publishers, 1981).

25. Lisa A. Parr and Bridget M. Waller, "Understanding Chimpanzee Facial Expression: Insights into the Evolution of Communication," *Social Cognitive and Affective Neuroscience* 1, no. 3 (2006): 221–228.

26. Paul Ekman and Erika L. Rosenberg, *What the Face Reveals: Basic and Applied Studies of Spontaneous Expression Using the Facial Action Coding System (FACS)* (New York: Oxford University Press, 1997).

27. Sarah-Jane Vick, Bridget M. Waller, Lisa A. Parr, Marcia C. Smith Pasqualini, and Kim A. Bard, "A Cross-species Comparison of Facial Morphology and Movement in Humans and Chimpanzees Using the Facial Action Coding System (FACS)," *Journal of Nonverbal Behavior* 31, no. 1 (2007): 1–20.

28. Satoshi Hirata, Goh Matsuda, Ari Ueno, Hirokata Fukushima, Koki Fuwa, Keiko Sugama, Kiyo Kusunoki, Masaki Tomonaga, Kazuo Hiraki, and Toshikazu Hasegawa, "Brain Response to Affective Pictures in the Chimpanzee," *Scientific Reports* 3, no. 1342 (2013).

29. Miho Nagasawa, Kensuke Murai, Kazutaka Mogi, and Takefumi Kikusui, "Dogs Can Discriminate Human Smiling Faces from Blank Expressions," *Animal Cognition* 14, no. 4 (2011): 525–533.

30. Tina Bloom and Harris Friedman, "Classifying Dogs' (Canis familiaris) Facial Expressions from Photographs," *Behavioural Processes* 96 (2013): 1–10.

31. John D. Reynolds, Mark A. Colwell, and Fred Cooke, "Sexual Selection and Spring Arrival Times of Red-necked and Wilson's Phalaropes," *Behavioral Ecology and Sociobiology* 18, no. 4 (1986): 303–310.

32. K. Yasukawa, J. L. Blank, and C. B. Patterson, "Song Repertoires and Sexual Selection in the Red-winged Blackbird," *Behavioral Ecology and Sociobiology* 7, no. 3 (1980): 233–238.

33. Myra O. Wiebe and M. Ross Lein, "Use of Song Types by Mountain Chickadees (Poecile gambeli)," *Wilson Bulletin* 111 no. 3 (1999): 368–375.

34. For a comprehensive review, see Clive K. Catchpole and Peter J. B. Slater, *Bird Song: Biological Themes and Variations* (Cambridge: Cambridge University Press, 1995).

35. Masakazu Konishi, "Development of Auditory Neuronal Responses in Avian Embryos," *Proceedings of the National Academy of Sciences* 70, no. 6 (1973): 1795–1798.

36. Peter Marler, "The Voice of the Chaffinch and Its Function as a Language," *Ibis* 98, no. 2 (1956): 231–261.

37. Ryo Ito and Akira Mori, "Vigilance Against Predators Induced by Eavesdropping on Heterospecific Alarm Calls in a Non-vocal Lizard Oplurus cuvieri cuvieri (Reptilia: Iguania)," *Proceedings of the Royal Society of London B: Biological Sciences* 277, no. 1685 (2010): 1275–1280.

38. Patricia M. Gray, Bernie Krause, Jelle Atema, Roger Payne, Carol Krumhansl, and Luis Baptista, "Biology and Music: The Music of Nature and the Nature of Music," *Science-International Edition-AAAS 291*, no. 5501 (2001): 52–53.

39. David A. Helweg, Adam S. Frankel, Joseph R. Mobley Jr., and Louis M. Herman. "Humpback Whale Song: Our Current Understanding," in *Marine Mammal Sensory*

Systems, ed. Jeanette A. Thomas, Ronald A. Kastelein, and Alexander A Supin (New York: Springer, 1992), 459–483.

40. Mark A. McDonald, Sarah L. Mesnick, and John A. Hildebrand, "Biogeographic Characterization of Blue Whale Song Worldwide: Using Song to Identify Populations," *Journal of Cetacean Research and Management* 8, no. 1 (2006): 55–65.

41. Constantine Nicholas Slobodchikoff, Bianca S. Perla, and Jennifer L. Verdolin, *Prairie Dogs: Communication and Community in an Animal Society* (Cambridge, Mass.: Harvard University Press, 2009).

42. Constantine Nicholas Slobodchikoff, C. Fischer, and J. Shapiro. "Predator-specific Words in Prairie Dog Alarm Calls," *American Zoologist* 26, no.105 (1986): 557.

43. Constantine Nicholas Slobodchikoff, Judith Kiriazis, C. Fischer, and E. Creef, "Semantic Information Distinguishing Individual Predators in the Alarm Calls of Gunnison's Prairie Dogs," *Animal Behaviour* 42, no. 5 (1991): 713–719.

44. Slobodchikoff, Perla, and Verdolin, *Prairie Dogs*.

45. Constantine Nicholas Slobodchikoff and R. Coast, "Dialects in the Alarm Calls of Prairie Dogs," *Behavioral Ecology and Sociobiology* 7, no. 1 (1980): 49–53.

46. Constantine Nicholas Slobodchikoff, *Chasing Doctor Dolittle: Learning the Language of Animals* (New York: Macmillan, 2012).

47. Robert M. Seyfarth, Dorothy L. Cheney, and Peter Marler, "Vervet Monkey Alarm Calls: Semantic Communication in a Free-ranging Primate," *Animal Behaviour* 28, no. 4 (1980): 1070–1094.

48. Robert M. Seyfarth, Dorothy L. Cheney, and Peter Marler, "Monkey Responses to Three Different Alarm Calls: Evidence of Predator Classification and Semantic Communication," *Science* 210, no. 4471 (1980): 801–803.

49. Robert M. Seyfarth and Dorothy L. Cheney, "Production, Usage, and Comprehension in Animal Vocalizations," *Brain and Language* 115, no. 1 (2010): 92–100.

50. Steven Pinker, *The Language Instinct: The New Science of Language and Mind* (London: Penguin, 1995).

51. Dorothy L. Cheney and Robert M. Seyfarth, "Vocal Recognition in Free-ranging Vervet Monkeys," *Animal Behaviour* 28, no. 2 (1980): 362–367.

52. Daniel Y. Takahashi, Darshana Z. Narayanan, and Asif A. Ghazanfar, "Coupled Oscillator Dynamics of Vocal Turn-taking in Monkeys," *Current Biology*: CB 23, no. 21 (2013): 2162–2168.

53. Roberta Salmi, Kurt Hammerschmidt, and Diane M. Doran-Sheehy, "Western Gorilla Vocal Repertoire and Contextual Use of Vocalizations," *Ethology* 119, no. 10 (2013): 831–847.

54. Christel Schneider, Josep Call, and Katja Liebal, "Do Bonobos Say NO By Shaking Their Head?," *Primates* 51, no. 3 (2010): 199–202.

55. Amy S. Pollick and Frans B. M. De Waal, "Ape Gestures and Language Evolution," *Proceedings of the National Academy of Sciences* 104, no. 19 (May 8, 2007): 8184–8189.

56. Mark E. Laidre, "Meaningful Gesture in Monkeys? Investigating Whether Mandrills Create Social Culture," *PloS one* 6, no. 2 (2011): e14610.

57. Marina Davila Ross, Michael J. Owren, and Elke Zimmermann, "The Evolution of Laughter in Great Apes and Humans," *Communicative & Integrative Biology* 3, no. 2 (2010): 191–194.

58. Dean Falk, "Comparative Anatomy of the Larynx in Man and the Chimpanzee: Implications for Language in Neanderthal," *American Journal of Physical Anthropology* 43, no. 1 (1975): 123–132.

59. Herbert S. Terrace, Laura-Ann Petitto, Richard J. Sanders, and Thomas G. Bever, "Can an Ape Create a Sentence?," *Science* 206, no. 4421 (1979): 891–902.

60. Sue Savage-Rumbaugh and Roger Lewin, *Kanzi: The Ape at the Brink of the Human Mind* (New York: Wiley, 1994).

61. Sue Savage-Rumbaugh, Kelly McDonald, Rose A. Sevcik, William D. Hopkins, and Elizabeth Rubert, "Spontaneous Symbol Acquisition and Communicative Use by Pygmy Chimpanzees (Pan paniscus)," *Journal of Experimental Psychology: General* 115, no. 3 (1986): 211.

62. Paul Raffaele, "Speaking Bonobo," *Smithsonian Magazine* 37 (2006): 74.

63. R. Allen Gardner, Beatrix T. Gardner, and Thomas E. Van Cantfort, *Teaching Sign Language to Chimpanzees* (Albany: State University of New York Press, 1989).

64. James M. Donovan and H. Edwin Anderson, *Anthropology and Law* (Oxford: Berghahn Books, 2005).

65. Roger S. Fouts, Deborah H. Fouts, and Thomas E. Van Cantfort, "The Infant Loulis Learns Signs from Cross-fostered Chimpanzees," in Gardner, Gardner, and Van Cantfort, *Teaching Sign Language to Chimpanzees*, 280–292.

66. Catherine Hobaiter and Richard W. Byrne, "The Gestural Repertoire of the Wild Chimpanzee," *Animal Cognition* 14, no. 5 (2011): 745–767.

67. Jana M. Iverson and Susan Goldin-Meadow, "What's Communication Got to Do with It? Gesture in Children Blind from Birth," *Developmental Psychology* 33, no. 3 (1997): 453.

EPILOGUE

1. Robert W. Mitchell, "Inner Experience as Perception(like) with Attitude," in *Experiencing Animals: Encounters Between Human and Animal Minds*, ed. Julie Ann Smith and Robert W. Mitchell (New York: Columbia University Press, 2012), 154–169.

2. Arii Watanabe, Uri Grodzinski, and Nicola S. Clayton, "Western Scrub-Jays Allocate Longer Observation Time to More Valuable Information," *Animal Cognition* 17, no. 4 (2014): 859–867.

3. Allison L. Foote and Jonathon D. Crystal, "Metacognition in the Rat," *Current Biology* 17, no. 6 (2007): 551–555.

4. Robert R. Hampton, "Rhesus Monkeys Know When They Remember," *Proceedings of the National Academy of Sciences* 98, no. 9 (2001): 5359–5362.

5. Benjamin M. Basile and Robert R. Hampton, "Monkeys Recall and Reproduce Simple Shapes from Memory," *Current Biology* 21, no. 9 (2011): 774–778.

6. Philip N. Johnson-Laird, *Mental Models: Towards a Cognitive Science of Language, Inference, and Consciousness* (Harvard University Press, 1983).

7. Steven Pinker, *How the Mind Works* (New York: Norton, 1997).

8. Robert W. Mitchell, "Mental Models of Mirror-self-recognition: Two Theories," *New Ideas in Psychology* 11, no. 3 (1993): 295–325.

9. Marc Bekoff and Paul W. Sherman, "Reflections on Animal Selves," *Trends in Ecology & Evolution* 19, no. 4 (2004): 176–180.

INDEX